Advances in Industrial Control

Other titles published in this series:

Digital Controller Implementation and Fragility
Robert S.H. Istepanian and James F. Whidborne (Eds.)

Optimisation of Industrial Processes at Supervisory Level
Doris Sáez, Aldo Cipriano and Andrzej W. Ordys

Robust Control of Diesel Ship Propulsion
Nikolaos Xiros

Hydraulic Servo-systems
Mohieddine Jelali and Andreas Kroll

Model-based Fault Diagnosis in Dynamic Systems Using Identification Techniques
Silvio Simani, Cesare Fantuzzi and Ron J. Patton

Strategies for Feedback Linearisation
Freddy Garces, Victor M. Becerra, Chandrasekhar Kambhampati and Kevin Warwick

Robust Autonomous Guidance
Alberto Isidori, Lorenzo Marconi and Andrea Serrani

Dynamic Modelling of Gas Turbines
Gennady G. Kulikov and Haydn A. Thompson (Eds.)

Control of Fuel Cell Power Systems
Jay T. Pukrushpan, Anna G. Stefanopoulou and Huei Peng

Fuzzy Logic, Identification and Predictive Control
Jairo Espinosa, Joos Vandewalle and Vincent Wertz

Optimal Real-time Control of Sewer Networks
Magdalene Marinaki and Markos Papageorgiou

Process Modelling for Control
Benoît Codrons

Computational Intelligence in Time Series Forecasting
Ajoy K. Palit and Dobrivoje Popovic

Modelling and Control of Mini-Flying Machines
Pedro Castillo, Rogelio Lozano and Alejandro Dzul

Ship Motion Control
Tristan Perez

Hard Disk Drive Servo Systems (2nd Ed.)
Ben M. Chen, Tong H. Lee, Kemao Peng and Venkatakrishnan Venkataramanan

Measurement, Control, and Communication Using IEEE 1588
John C. Eidson

Piezoelectric Transducers for Vibration Control and Damping
S.O. Reza Moheimani and Andrew J. Fleming

Manufacturing Systems Control Design
Stjepan Bogdan, Frank L. Lewis, Zdenko Kovačić and José Mireles Jr.

Windup in Control
Peter Hippe

Nonlinear H_2/H_∞ Constrained Feedback Control
Murad Abu-Khalaf, Jie Huang and Frank L. Lewis

Practical Grey-box Process Identification
Torsten Bohlin

Control of Traffic Systems in Buildings
Sandor Markon, Hajime Kita, Hiroshi Kise and Thomas Bartz-Beielstein

Wind Turbine Control Systems
Fernando D. Bianchi, Hernán De Battista and Ricardo J. Mantz

Advanced Fuzzy Logic Technologies in Industrial Applications
Ying Bai, Hanqi Zhuang and Dali Wang (Eds.)

Practical PID Control
Antonio Visioli

(continued after Index)

Alberto Cavallo · Giuseppe De Maria
Ciro Natale · Salvatore Pirozzi

Active Control of Flexible Structures

From Modeling to Implementation

Springer

Prof. Alberto Cavallo
Prof. Giuseppe De Maria
Prof. Ciro Natale
Dr. Salvatore Pirozzi

Seconda Università di Napoli
Fac. Ingegneria
Dipto. Ingegneria dell'Informazione
Via Roma 29
81031 Aversa Caserta
Italy

alberto.cavallo@unina2.it
giuseppe.demaria@unina2.it
ciro.natale@unina2.it
salvatore.pirozzi@unina2.it

ISSN 1430-9491
ISBN 978-1-4471-5707-6 ISBN 978-1-84996-281-0 (eBook)
DOI 10.1007/978-1-84996-281-0
Springer London Dordrecht Heidelberg New York

British Library Cataloguing in Publication Data
A catalogue record for this book is available from the British Library

Cover design: eStudioCalamar, Figueres/Berlin

Printed on acid-free paper

Springer is part of Springer Science+Business Media (www.springer.com)

Advances in Industrial Control

Professor (Emeritus) O.P. Malik
Department of Electrical and Computer Engineering
University of Calgary
2500, University Drive, NW
Calgary, Alberta
T2N 1N4
Canada

Professor K.-F. Man
Electronic Engineering Department
City University of Hong Kong
Tat Chee Avenue
Kowloon
Hong Kong

Professor G. Olsson
Department of Industrial Electrical Engineering and Automation
Lund Institute of Technology
Box 118
S-221 00 Lund
Sweden

Professor A. Ray
Department of Mechanical Engineering
Pennsylvania State University
0329 Reber Building
University Park
PA 16802
USA

Professor D.E. Seborg
Chemical Engineering
3335 Engineering II
University of California Santa Barbara
Santa Barbara
CA 93106
USA

Doctor K.K. Tan
Department of Electrical and Computer Engineering
National University of Singapore
4 Engineering Drive 3
Singapore 117576

Professor I. Yamamoto
Department of Mechanical Systems and Environmental Engineering
The University of Kitakyushu
Faculty of Environmental Engineering
1-1, Hibikino,Wakamatsu-ku, Kitakyushu, Fukuoka, 808-0135
Japan

To our families

Series Editors' Foreword

The series *Advances in Industrial Control* aims to report and encourage technology transfer in control engineering. The rapid development of control technology has an impact on all areas of the control discipline. New theory, new controllers, actuators, sensors, new industrial processes, computer methods, new applications, new philosophies..., new challenges. Much of this development work resides in industrial reports, feasibility study papers and the reports of advanced collaborative projects. The series offers an opportunity for researchers to present an extended exposition of such new work in all aspects of industrial control for wider and rapid dissemination.

The widespread presence of flexible structures in urban and industrial infrastructure and transportation systems is easily overlooked as we perform the everyday tasks of modern living. While civil, mechanical, aeronautical and space engineers have become more adventurous in their designs for buildings, bridges, rail transport, automobiles, aircraft and space structures (to name just a few flexible structure applications) so have these designs have had to accommodate the potential for, and the phenomena of increased structural flexibility and vibration. Large abrupt deformations can cause the structural integrity of many designs to fail. One response to this is to develop designs that are able to withstand such deformations, as is found in some earthquake-prone countries where buildings and bridges can be designed to survive moderate forces of nature. An alternative is to use techniques that ensure that flexible structures operate away from conditions that can cause destruction, or that eliminate any conditions that could lead to more catastrophic situations. The latter approach is used, for example, in transportation systems.

Control engineers have had a long-standing interest in the control of flexible structures since to avoid weight penalties, designers have wanted to use slender arms and thinner plates for applications in robotics (flexible robotic arms), aeronautics (fixed wings and rotary blades) and in space (extendable solar panels and antennae). Flexible structures are examples of distributed-parameter systems described by models that use partial differential equations to capture the spatiotemporal dependence of quantities and properties in the system. These systems can be assumed to be subject to linear vibratory motions about a stable equilibrium condition, and this type of linear analysis leads to the flexible structure having one or more resonant frequencies with associated modal responses. This initiates an important requirement when finding finite-dimensional lumped parameter models based on distributed-parameter model idealisations. It is also the starting point for *Active Control of Flexible Structures: From Modelling to Implementation* by Alberto Cavallo, Giuseppe De Maria, Ciro Natale, Salvatore Pirozzi.

The chapters of the monograph follow the time-honoured path to developing some novel control solutions for flexible-structure-control problems. Modelling and identification open the monograph in Chapters 2 and 3. There is an extended development of H_∞ and H_2 control solutions presented in Chapter 4. This particular chapter gives a fascinating in-depth presentation of the analytical twists and turns that modify advanced robust control solutions for flexible structure applications. The final three chapters (5, 6, and 7) cover implementation, experimental results and future research directions. The test rig experiments of Chapter 6 use a fuselage skin panel from a Boeing 717 showing just how close this work is to real applications.

The disciplines impinging on flexible-structure-control problems include the aeronautical, space, mechanical, robotics, mechatronics and control engineering fields, consequently a wide readership is envisaged for this well-focussed addition to the *Advances in Industrial Control* series of monographs.

Industrial Control Centre
Glasgow
Scotland, UK
2010

M.J. Grimble
M.A. Johnson

Acknowledgements

Thanking all the people who made possible this publication without forgetting someone is a difficult task, especially when so many people deserve a grateful acknowledgement.

First of all, we have to recall Leonardo Lecce (Professor of Aeroelasticity at Università di Napoli Federico II) who invited Giuseppe De Maria and Alberto Cavallo more than ten years ago to participate in a European project dealing with active noise and vibration control through magnetostrictive actuators, MESA, which was coordinated by Chris May (Universität des Saarlandes). He gave us the opportunity to exploit our background on control theory for tackling the challenging problem of vibration control of aeronautic structures. The successful conclusion of the MESA project led to a second research project on magnetostrictive actuators for vibration control, with even more challenging objectives, MESEMA, which was coordinated by Leonardo Lecce. During the project the authors had to tackle the really difficult problem of noise and vibration control in a broadband disturbance field. The strong interdisciplinary nature of the problem required a shoulder to shoulder collaboration with brilliant aeronautic engineers like Ernesto Monaco and Francesco Franco, who introduced us to the fascinating field of the modelling of flexible structures. They taught us a language that was new for all of us, that of structural engineers and, at the same time, they had to learn a language that was new for them, that of control engineers. On the topic of the modelling of flexible structures, we cannot forget to mention our colleague Vincenzo Minutolo (Professor of Solid and Structural Mechanics at Seconda Università di Napoli), who had the patience to explain to us much of the basics of continuum mechanics.

Once again, the successful results of the MESEMA project would have never been achieved without the effort of many people, who we want to warmly thank. Among many others, Luigi Zeni and Aldo Minardo made an important contribution in the development of the sensing systems of the smart auxiliary dampers. This miracle of mechanics came out of an original idea of Chris May, who is now a great friend of all of us. Collaborating with him is always very stimulating, especially because his original devices always lead to nonlinear systems, which test our control design skills.

Finally, we warmly thank Pietro Pagliarulo for his contribution to the dynamic model of the smart auxiliary damper and Michele Iadevaia for providing us the experimental data of the modal analysis carried out in Section 3.6.2.

Aversa, Italy
March 2010

Alberto Cavallo
Giuseppe De Maria
Ciro Natale
Salvatore Pirozzi

Contents

1 Introduction 1

2 Modelling of Flexible Structures 7
2.1 Introduction 7
2.2 Preliminary Definitions 8
2.3 The Boundary Value Problem 9
2.3.1 Uniform String 13
2.3.2 Uniform Cantilever Beam 15
2.4 Approximate Finite-dimensional Models 17
2.4.1 Damped Elastic Structures 20
2.4.2 Finite-dimensional Systems 22
2.5 Conclusions 27
References 27

3 Identification of Flexible Structures 29
3.1 Introduction 29
3.2 Basics of Subspace-based Identification 31
3.2.1 Time Domain Approach 32
3.2.2 Frequency Domain Approach 34
3.3 Identification Problem for a Flexible Structure 38
3.4 Identification Procedure 40
3.4.1 First Stage 40
3.4.2 Second Stage 41
3.4.3 Identification Procedure with Non-colocated Measurements 45
3.5 Application to Experimental Modal Analysis 46
3.6 Illustrative Examples 48
3.6.1 Identification of a Model for Control 48
3.6.2 Identification for Experimental Modal Analysis 51
3.7 Conclusions 53
References 53

4 Modal Feedback Control of Flexible Structures ... 55
4.1 Introduction ... 55
4.2 Spillover Problem ... 58
4.3 Internal Model Control ... 61
4.3.1 Basic Approach ... 62
4.3.2 Adaptive Approach ... 63
4.4 Modal Control ... 67
4.4.1 Case Study: State Feedback ... 72
4.4.2 State Observer ... 74
4.5 Conclusions ... 76
References ... 77

5 Optimal Feedback Control of Flexible Structures ... 79
5.1 Introduction ... 79
5.2 Control Model ... 81
5.3 Choice of the Disturbance Matrix ... 84
5.4 Bandpass Controllers ... 84
5.5 $\mathscr{H}_\infty$ Controller Design ... 86
5.6 $\mathscr{H}_2$ Controller Design ... 91
5.7 Numerical Example ... 92
5.8 Further Considerations on the $\mathscr{H}_2$ Controller ... 94
5.9 Optimal Scaling Matrix Selection ... 98
5.9.1 Formulation of the Optimsation Problem ... 98
5.9.2 Scaling Matrix Selection Example ... 101
5.10 Selection of Individual Modes ... 104
5.10.1 Mode Selection and Non-colocated Feedback Example ... 108
5.10.2 Computing the Gramians ... 113
5.11 Conclusions ... 120
References ... 121

6 Implementation Issues ... 123
6.1 Introduction ... 123
6.2 Conventional Actuators ... 124
6.3 Smart Material-based Devices ... 126
6.3.1 Piezoelectric Ceramics ... 127
6.3.2 Magnetostrictive Materials ... 128
6.4 Effects of Hysteresis ... 129
6.4.1 Hysteresis Modelling ... 130
6.4.2 Hysteresis Compensation: Methodology ... 132
6.4.3 Hysteresis Compensation: Experimental Results ... 133
6.5 A Smart Auxiliary Mass Damper ... 134
6.5.1 Actuator Description and Modelling ... 136
6.5.2 The Optical Sensor ... 139
6.5.3 The Control System ... 141
6.5.4 Experimental Results of Actuator Control ... 143

6.6 Self-sensing Control . . . 144
6.7 The dSPACE Platform for Rapid Control Prototyping . . . 146
6.7.1 Processor Boards . . . 147
6.7.2 I/O Boards . . . 148
6.7.3 Software Development Environment . . . 148
6.8 Conclusions . . . 149
References . . . 150

7 Experimental Results . . . 153
7.1 Introduction . . . 153
7.2 Identification of the Dynamic Model of the Panel . . . 155
7.3 Active Control of Vibrations with Piezo Actuators . . . 163
7.3.1 Analytically Designed Controllers . . . 166
7.3.2 Numerically Designed Controller Using LMI . . . 168
7.4 Active Vibration Control with a Smart Auxiliary Mass Damper . . . 176
7.5 Conclusions . . . 180
References . . . 181

8 Conclusions and Future Developments . . . 183
8.1 Conclusions . . . 183
8.2 Future Developments . . . 185

Appendix Dynamic Model of the Aeronautical Panel . . . 187

Index . . . 195

Acronyms

ARE	Algebraic Riccati Equation
DIDO	Differential Input Differential Output
DFB	Distributed Feedback
DFT	Discrete Fourier Transform
DOF	Degree of Freedom
EU	European Union
FBG	Fiber Bragg Grating
FEM	Finite Element Model
FFT	Fast Fourier Transform
FRF	Frequency Response Function
IDFT	Inverse Discrete Fourier Transform
LFT	Linear Fractional Transformation
LMI	Linear Matrix Inequality
LQG	Linear Quadratic Gaussian
LTI	Linear Time Invariant
LTR	Loop Transfer Recovery
MDOF	Multidegree of Freedom
MIMO	Multi-input Multi-output
NOL	Naval Ordnance Laboratory
NRL	Naval Research Laboratory
PLL	Phase Locked Loop
PM	Preisach Model
PDF	Preisach Distribution Function
PZT	Piezoelectric Zirconate Titanate
RCP	Rapid Control Prototyping
RTI	Real-time Interface
SISO	Single-input Single-output
SVD	Singular Value Decomposition
TITO	Two-input Two-output

Chapter 1
Introduction

The subject of the book is the design of control systems for flexible structures. Within the control engineering community the term *flexible structure* is most commonly used to refer to a body or a system of interconnected bodies subject to a vibratory motion, in which the system oscillates about a stable equilibrium position in *free* or *forced* conditions. In this book, attention is focused on so-called *linear vibration* to denote situations in which the vibrating system can be modelled as a linear one. Generally speaking, small amplitudes of motion are the range where linearity holds. A *free vibration* is characterised by a repetitive motion of the system about its static equilibrium position; the motion is performed under the action of forces inherent in the system itself, the so-called *restoring forces*. *Forced vibration* is a vibratory motion imposed on the system by the action of an external excitation. One of the specific features of flexible structures in motion is *resonance*, which is a phenomenon characterised by strong amplification of the motion at a specific frequency called the *natural frequency*; there can be several frequencies that the structure resonates at. The movement of a structure at these frequencies is harmonic, or sinusoidal, and remains at the same pattern of deformations, called *mode shape*. The mode shapes are uncoupled and, being independent, they can be excited separately. The total structural response is the sum of responses of individual modes. This property is of fundamental importance in tackling the control problem of flexible structures.

The research work of the control community in the field of the control of flexible structures is devoted to designing control systems that are able to suppress or to reduce the amplitude of unwanted vibration in complex systems. Many applications are of great interest to control engineers, generally concerning mechanical structures, such as space structures, industrial and space robotics, and earth and air transportation systems. Structural vibration and elasticity in the joint of a robotic manipulator are the origin of many problems in fine positioning systems. Elastic joints and/or flexible arms produce oscillation about the final set point and a lack of precision with respect to the final set point designed. On the other hand, there is an inherent difficulty of simultaneously carrying out structural vibration suppression and precision manoeuvres and/or pointing the tip of the manipulator at an assigned position. That means structural vibrations impose a strong requirement on the bandwidth of the controlled system. The same problem arises with the attitude control of space structures, where the *per* unit mass cost of placing these systems into orbit often leads to extremely flexible (and light) structural designs with many modes in the controller bandwidth. In earth and air transportation systems, vibrations impose strong requirements on structural design and on the design of the control systems concerning ride/flight safety, the integrity of the structure and the comfort of the pilots and the passengers. Such requirements are more severe for aircraft and helicopters. With respect to flight safety, for example, one of the main problems concerning control surfaces is flutter instability, which is the unstable oscillation caused by interaction between aerodynamic forces, elastic reactions in the structure and the force of inertia. Unless damped quickly, flutter will break a wing of an aircraft or a blade of a helicopter. The structural design cure for flutter is usually to dispose the weight so that the centre of gravity of the wing is as far forward as possible. Ailerons, elevators and rudders are mass-balanced to prevent flutter. It can be said that a curtain was raised from the flutter instability of structures by Tacoma Narrows Bridge failure in 1940. The active suppression of flutter is still an open research topic.

A different research topic is related to structural fatigue produced by structural vibrations, which can lead to damage. In the case of aeronautical structures this problem imposes the use of a health monitoring system, especially for fuselage components. Nowadays aircraft design is based on the *damage tolerance approach*. According to this method structural damage can occur but must be monitored, for example, in terms of crack propagation speed, in order to avoid dangerous structural breaks. For this reason, some key components are doubled in aircraft. The damage tolerance philosophy requires an accurate health monitoring program and it introduces some weight penalties. A significant research effort has been pursued in the last years in the development of health monitoring algorithms for aeronautical systems, most of them based on the frequency response functions of the monitored structure.

Structural vibrations, and the consequently produced acoustic noise in transportation systems, have a strong impact on the comfort of the pilots and passengers. They produce physical stress, which, in the case of pilots, concerns ride safety also. As regards earth transportation systems such as trains, structural vibrations producing relevant noise inside the carriage are mainly due to the wheel–railtrack impact and passive dampers are used to reduce the effects, in terms of vibration and acoustic noise, in the carriage. Vibrations produced by the interactions of aerodynamic forces with the structure are taken into account in the design of trains, especially in the case of high speed trains. With respect to automotive systems, structural vibrations are certainly caused by the quality of the road surface. The use of dampers reduces the influence of the impact on the passenger compartment. The new generation of cars makes use of active controlled dampers. However, the main source of vibration and noise in automotive systems is the engine, and the cure is a passive isolation of the passenger compartment, which results in a weight penalty.

In air transportation systems, both fixed-wing aircraft and rotorcraft are characterised by a number of very demanding requirements: the passenger ride comfort should be increased, the physical stress of pilots should be decreased, a less severe vibratory environment for equipment should be the goal, there should be less structural fatigue problems, ride in helicopters should be jet-smooth. The major sources of structural vibrations and the nature of vibrations are different in fixed-wing aircraft and rotorcraft; with reference to fixed-wing aircraft it depends on the propeller. The major source of vibrations and noise in turbo-fan aircraft can be imputed to the so-called *bump noise*, a typical disturbance affecting cruise flight. This is due to the dynamic interaction between the fuselage and the external air flow, which results in a broadband vibratory field on the aircraft structure and produces a broadband acoustic noise field inside the cabin. In turbo-prop aircraft the major source of vibrations is the dynamic interaction of the air flow produced by the blades with the fuselage mainly in the blade section of the fuselage itself, and this results in a tonal vibration and acoustic noise field. Vibratory loads in helicopters arise from a variety of sources such as the main rotor system, the aerodynamic interaction between the rotor and the fuselage, the tail rotor, the engine and transmission. However, the most significant source of vibrations in a helicopter is the main rotor because of the unsteady aerodynamic environment acting on highly flexible rotating blades. The vibratory and noise field can be assumed to be tonal. The reduction of vibration levels in helicopters below acceptable limits is one of the central problems facing rotorcraft designers. In order to comply with the requirements concerning structural vibrations and acoustic noise inside the fuselage of an aircraft, concentrated research has been developing to design noise and vibration control systems, passive ones or active ones. Concerning the suppression of structural vibrations in fixed wing aircraft at high frequencies, passive distributed damping materials are attached on the skin panels of the fuselage. However, to reduce the interior noise caused by vibrations in the acoustic frequency range, the strategy, in the case of tonal noise field, is twofold. The first approach uses supplementary noise sources, while the second exploits vibration absorbers mounted on the structure, which means acting directly on the noise source.

Historically, the birth of active noise reduction by using automatic control approaches and electronic technologies dates back to 1936, with the patent by Lueg: "Processing of silencing sound oscillations". The idea is still the basis for feedforward noise control: one microphone senses the sound produced in a duct by a noise source, a controller (a simple electronic amplifier) processes the signal from the microphone, by amplifying it and shifting its phase by a suitable angle, then feeds a loudspeaker placed in the duct downstream from the noise wave (see Figure 1.1). If the control gain and shift are suitably chosen, the loudspeaker produces a "secondary wave" (dashed waveform in the figure) that cancels the noise. The above approach is simple and effective, but suffers from several drawbacks, mainly due to the lack of *robustness*; modelling errors or parameter variations in the path from the microphone to the loudspeaker may even lead to vibration amplification, as the two waves (disturbance and control) may undergo constructive instead of destructive in-

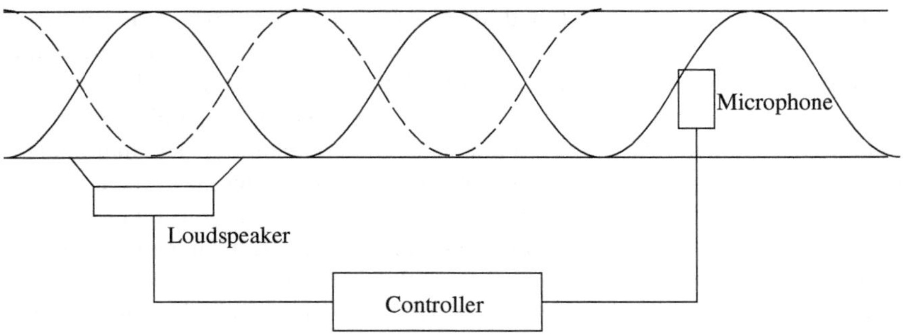

Figure 1.1 Lueg's patent

ference, *i.e.*, they may sum instead of subtract. Many studies have originated from this observation, resulting in a more systematic approach to the problem.

The alternative strategy to control the noise source directly has been successfully implemented in turbo-prop aircraft and is nowadays widely used. It consists of mounting resonant devices tuned at the frequency of the vibrations on the bulkhead of the fuselage, which are the source of the acoustic noise. They are *passive dynamic vibration absorbers* and the major drawback is the loss of the tuning when there is a change in the vibration frequencies. In turbo-fan aircraft a broadband vibratory and noise field is involved, typically in the range 300–1000 Hz, and passive devices and strategies are not useful to alleviate the effects, in terms of both acoustic noise and structural vibrations. To solve this problem, researchers have been producing a significant effort in the field of active noise and vibration control of flexible structures for several years, developing research projects where the leading aeronautical industries in the world are involved.

The authors of this book were involved in the last ten years in two projects financed by the European Community in the 5th and 6th Framework programmes. In the latter, the MESEMA project (magnetoelastic energy systems for even more electric aircraft), an entire work package was devoted to active noise and vibration control and contained two development activities. One aimed at improving the cabin environment in turbo-fan aircraft by reducing noise levels through the application of magnetostrictive devices, as active dynamic vibration absorbers, on the structural members. The second one aimed at the development of smart force generators for combating vibrations in the cockpit of helicopters. Some leading industries, including ALENIA Aeronautica, EUROCOPTER Deutschland and ZF-Luftfahrttechnik, were involved in this work package along with European research centres. Innovative modelling and identification procedures as well as innovative active control strategies were developed by the authors during the project and tested both on single structural members and on a full-scale fuselage segment of a real aircraft. The successful results of the project encouraged further research effort of the authors in the field of active control of flexible structures. This book contains the recent results achieved on the identification of continuous-time models of flexible structures and on active control methodologies for flexible structures, as well as exploitation of innovative actuation and sensing technologies. The book is organised as follows.

Chapter 2 is devoted to illustrating the basics of the modelling of flexible structures. Assuming that the equation of motion of a flexible structure can be formulated as a boundary value problem, the eigensolutions are found *via* modal analysis by solving a differential eigenvalue problem. A modal truncation procedure is presented in order to obtain an approximate finite dimensional model of the flexible structure.

Chapter 3 describes an identification procedure specifically devised to estimate the parameters of the flexible structure model based on modal coordinates. The procedure consists of a non-iterative algorithm that identifies a continuous-time state space model *via* a grey-box method in the frequency domain based on a subspace approach. The first part of the chapter presents a review of the existing identification methodologies most commonly used to identify dynamic models of flexible structures. A numerical case study is carried out to illustrate the features of the algorithm and the results of experimental modal analysis, based on the identification procedure of a beam are also presented.

Chapters 4 and 5 are devoted to the control of vibrating systems. The control goal is to reduce the effect of persistent disturbances on the structure by using velocity measurements obtained by integrating signals from accelerometers suitably placed on the structure, and smart actuators exciting the system in such a way to create vibrations able to counteract the effect of the disturbance. From the disturbance point of view, two shapes of the signal are considered, namely tonal and broadband disturbances. The distinction between systems with low and high model density is also outlined. Specifically, in Chapter 4, internal model strategies are considered for systems with tonal disturbances and low modal densities. In Chapter 5, optimal approaches with $\mathscr{H}_2$ and $\mathscr{H}_\infty$ cost indices are considered for high modal density structures, even in the most complex broadband disturbance case. Even though most control approaches rely on the assumption that the couples of actuators and sensors used for feedback control are located in the same points, *i.e.*, they are *colocated*, by using a $\mathscr{H}_2$ cost index a non-colocated control strategy is also illustrated, namely additional sensors are placed where no actuator is present.

Chapter 6 provides technical information that is useful in choosing the type of actuator and/or sensor for the practical realisation of an active vibration control system and presents some techniques that allow optimising performance by properly using non-conventional devices. The most interesting of such devices for application in the control of flexible structures are the piezoelectric ceramics and magnetostrictive actuators. A rapid prototyping control system, which is fundamental to quickly testing and tuning various control strategies on a real time system, is also presented.

Chapter 7 addresses the problem of reducing the vibrations of a real flexible structure subjected to inaccessible broadband disturbances. The identification procedure of continuous-time models of flexible structures presented in Chapter 3 and the active control strategies presented in Chapter 5 are experimentally tested on an aeronautical structure, *i.e.*, a stiffened skin panel of a BOEING 717.

Chapter 2
Modelling of Flexible Structures

2.1 Introduction

The basic skills required for feedback control engineering are the ability to model dynamic systems and to analyse dynamic responses. Generally speaking, the modelling process can be viewed as the first step involved in control system design. Once a model is obtained in the form of dynamic equations, the next step is to solve the equations to discover the nature of the response of the system.

It should be pointed out that models represent an idealisation of the actual physical systems, since they are based on a set of assumptions and have limits of validity that must be specified at some point of the investigation. So, for the same system it is possible to construct a number of models, the "best" being the simplest one that retain all the essential features of the actual system under study with reference to the objectives of the investigation. The modelling phase, in this monograph, is devoted to the analysis and prediction of vibratory motion of physical systems, in which the system under study oscillates about a stable equilibrium position as a consequence of a perturbing disturbance which, in turn, starts the motion by displacing the system from such a position. Attention will focus on "linear vibration" or, more precisely, to situations in which the vibrating systems can be modelled as a linear one. In general, small amplitudes of motion are the range where linearity holds and the cornerstone of linearity is the principle of superposition.

Dynamic systems, according to their physical characteristics, can be *continuous* or *discrete*. Flexible or elastic structures are continuous systems since their mass and stiffness are distributed in space and represented by functions of spatial variables. The motion of a given flexible structure is governed by partial differential equations to be satisfied over the domain of extension of the structure and boundary conditions to be satisfied at the boundaries of the domain. In general, closed-form solutions to the partial differential equations describing distributed structures are not feasible. Hence, it is useful and advisable to replace distributed characteristics with discrete ones to produce approximate solutions by resorting to some discretisation procedures and to describe the motion of the flexible structure under study by means of a finite set of ordinary differential equations. Discrete parameter systems have finite degrees-of-freedom (DOF), *i.e.*, only a finite number of independent coordinates is needed to define their motion. The widely used assumed-modes or finite-element methods are, in essence, spatial discretisation procedures that retain aspects of either continuous or discrete systems. Whatever discretisation method is used, one advantage is the possibility to improve the accuracy of the analysis by increasing the number of DOF. Continuous systems can be seen as limiting cases of discrete ones. The presentation of the methodologies to formulate dynamic equations and boundary conditions, the so-called *boundary value problem*, for distributed parameter systems, as well as the presentation of an extensive review of the discretisation procedures are beyond the scope of this chapter and the reader is referred to the specialised textbooks cited in the references. In this chapter the general boundary value problem for undamped and damped systems will be presented. The response will be obtained by modal analysis assuming that the solution of the differential eigenvalue problem, involved in the solution of the boundary value problem, is available. The procedure will be illustrated with the aim of investigating the nature of the eigensolutions in terms of natural frequencies and natural modes of oscillation. A spatial discretisation procedure will be presented in order to obtain a discrete or finite-dimensional model of the flexible structure, with the aim of identifying the structure of the model that will be helpful in the grey-box identification procedure, illustrated in Chapter 3, as well as in the control system design phase described in Chapters 4 and 5.

2.2 Preliminary Definitions

Let $\mathscr{W}$ be the vector space of square-integrable (in the Lebesgue sense) $\mathbb{R}^m$-valued functions defined on a domain $X \subseteq \mathbb{R}^l$. The expression

$$< f,g >_{\mathscr{W}} = \int_X f^{\mathrm{T}}(x)g(x)\mathrm{d}x, \quad x \in X,\ f,g \in \mathscr{W} \tag{2.1}$$

defines the inner product in $\mathscr{W}$, where the expression $f^{\mathrm{T}}(x)g(x)$, for a fixed $x \in X$, indicates the inner product in $\mathbb{R}^m$.

Let $\mathscr{U}$ and $\mathscr{V}$ be two inner-product vector spaces of square-integrable functions of any dimension defined on a domain $X \subseteq \mathbb{R}^l$ and let $\mathscr{A}$ be a linear operator $\mathscr{A}$: $\mathscr{U} \rightarrow \mathscr{V}$. The linear operator $\mathscr{A}^* : \mathscr{V} \rightarrow \mathscr{U}$ is said to be the *adjoint operator* of $\mathscr{A}$ if it is such that

$$< \mathscr{A} f,g >_{\mathscr{V}} = < f,\mathscr{A}^* g >_{\mathscr{U}}, \quad \forall f \in \mathscr{U}, \forall g \in \mathscr{V}. \tag{2.2}$$

If the adjoint operator $\mathscr{A}^*$ of $\mathscr{A}$ exists, it is unique. The linear operators for which the adjoint operator exists have the following properties:

(i) $(\mathscr{A}^*)^* = \mathscr{A}$,
(ii) if $\mathscr{A} : \mathscr{U} \rightarrow \mathscr{V}$ and $\mathscr{B} : \mathscr{V} \rightarrow \mathscr{U}$, then $(\mathscr{B} \circ \mathscr{A})^* = \mathscr{A}^* \circ \mathscr{B}^*$,

(iii) $(\mathscr{A}^{-1})^* = (\mathscr{A}^*)^{-1}$,
(iv) $\mathscr{A}^* \circ \mathscr{A} = (\mathscr{A}^* \circ \mathscr{A})^*$ and $\mathscr{A} \circ \mathscr{A}^* = (\mathscr{A} \circ \mathscr{A}^*)^*$.

A linear operator $\mathscr{L}$ for which the adjoint operator $\mathscr{L}^*$ exists and is such that $\mathscr{L} = \mathscr{L}^*$ is said to be *self-adjoint*. The linear operators $\mathscr{A}^* \circ \mathscr{A}$ and $\mathscr{A} \circ \mathscr{A}^*$ are self-adjoint.

Let u be a given function; the operator $\mathscr{L} : \mathscr{U} \to \mathscr{U}$ is said *positive semi-definite* if

$$< u, \mathscr{L}u >= \int_X u^{\mathrm{T}}(x)\mathscr{L}u(x)\mathrm{d}x \geq 0, \quad \forall u \in \mathscr{U}. \tag{2.3}$$

Consider two functions $f, g \in \mathscr{U}$ piecewise continuous in the domain $X \subseteq \mathbb{R}^l$, if

$$< f, g >= \int_\Pi f^{\mathrm{T}}(x)g(x)\mathrm{d}x = 0, \tag{2.4}$$

they are said to be *orthogonal* over the domain X.

Consider two functions $f, g \in \mathscr{U}$ piecewise continuous in the domain $X \subseteq \mathbb{R}^l$ and the operator $\mathscr{L} : \mathscr{U} \to \mathscr{U}$, if

$$< f, \mathscr{L}g >= \int_\Pi f^{\mathrm{T}}(x)\mathscr{L}g(x)\mathrm{d}x = 0, \tag{2.5}$$

they are said to be *orthogonal with respect to the operator* $\mathscr{L}$ over the domain X.

2.3 The Boundary Value Problem

The equation of motion of a flexible structure with mass distribution function $m(p)$ subject to a generalised force distribution $f(p,t)$, with $p \in \Pi \subseteq \mathbb{R}^l$ the position vector of a point of the structure and Π an open, bounded and regular domain, can be written as a general boundary value problem [2] by the following partial differential equation together with boundary and initial conditions

$$m(p)\ddot{w}(p,t) + \mathscr{L}w(p,t) = f(p,t), \qquad p \in \Pi \subseteq \mathbb{R}^l, \tag{2.6}$$

$$\mathscr{B}_i w(p,t)|_{p\in\partial\Pi} = 0, \qquad i = 1, 2, \ldots, h, \tag{2.7}$$

$$w(p,0) = w_0(p),\ \dot{w}(p,0) = w_1(p), \tag{2.8}$$

where $w(p,t) \in \mathbb{R}^s$ denotes the displacement from an equilibrium position of the point with coordinates p of the structure, $\mathscr{L}$, called *spatial differential stiffness operator*, is a linear, self-adjoint and positive semi-definite spatial differential operator of order $2h$; $\mathscr{B}_i$ are spatial differential operators of maximum order $2h-1$; $\partial\Pi$ denotes the boundary of the domain Π; $w_0(p)$ and $w_1(p)$ are given $\mathbb{R}^s$-valued functions representing the initial conditions.

Consider the free vibration problem, $f(p,t) = 0$, and assume a solution of Equation 2.6 in the form

$$w(p,t) = W(p)\eta(t), \tag{2.9}$$

where $W(p)$ is the $\mathbb{R}^s$-valued space-dependent function and $\eta(t)$ is the time-dependent function. Substituting Equation 2.9 into Equation 2.6 it is very easy to show that the boundary value problem 2.6–2.8 can be rewritten as in the following two equations:

$$\ddot{\eta}(t) = -\lambda\eta(t), \qquad \eta(0) = \eta_0,\ \dot{\eta}(0) = \eta_1, \tag{2.10}$$

$$\mathscr{L}W(p) = \lambda m(p)W(p), \qquad p \in \Pi, \tag{2.11}$$

where λ is a constant parameter to be determined and $W(p)$ must satisfy the boundary conditions, *i.e.*,

$$\mathscr{B}_i W(p)|_{p\in\partial\Pi} = 0, \qquad i = 1,2,\ldots,h. \tag{2.12}$$

The problem 2.11 and 2.12 is known as the *differential eigenvalue problem, i.e.*, the problem of looking for the values of parameter λ for which Equation 2.11 has nontrivial solutions satisfying the boundary conditions. The set of values of the parameter λ that satisfy Equation 2.11 and the boundary conditions 2.12 are called the characteristic values or *eigenvalues*, and the associated non-trivial solutions for the vector functions $W(p)$ are called the characteristic functions or *eigenfunctions*. The solution of the characteristic equation 2.11 consists of a countable infinite set of eigenvalues and the corresponding set of eigenfunctions is unique except for their scale factors. The assumption that the linear operator $\mathscr{L}$ is self-adjoint has some important consequences. The following proposition can be easily proved.

Proposition 2.1. *The eigenfunctions corresponding to distinct eigenvalues are orthogonal with respect to the mass distribution and the stiffness operator over the domain* Π.

Proof. Let W_j and W_k with $j \neq k$ two vector eigenfunctions, as solutions of the eigenvalue problem 2.11, then

$$\mathscr{L}W_j(p) = \lambda_j m(p)W_j(p), \qquad p \in \Pi, \tag{2.13}$$

$$\mathscr{L}W_k(p) = \lambda_k m(p)W_k(p), \qquad p \in \Pi. \tag{2.14}$$

Pre-multiplying Equation 2.13 by W_k^{T} and Equation 2.14 by W_j^{T} and integrating over the domain Π yields

$$\int_\Pi W_k^{\mathrm{T}}(p)\mathscr{L}W_j(p)\mathrm{d}p = \int_\Pi \lambda_j m(p)W_k^{\mathrm{T}}(p)W_j(p)\mathrm{d}p, \tag{2.15}$$

$$\int_\Pi W_j^{\mathrm{T}}(p)\mathscr{L}W_k(p)\mathrm{d}p = \int_\Pi \lambda_k m(p)W_j^{\mathrm{T}}(p)W_k(p)\mathrm{d}p. \tag{2.16}$$

Subtracting side by side and taking into account that $\mathscr{L}$ is a self-adjoint differential operator, the following result is obtained:

$$(\lambda_j - \lambda_k)\int_\Pi m(p)W_k^{\mathrm{T}}(p)W_j(p)\mathrm{d}p = 0, \quad \forall j \neq k. \tag{2.17}$$

Then, $\forall j \neq k$ for which $\lambda_j \neq \lambda_k$ Equation 2.17 becomes

$$\int_{\Pi} m(p) W_k^{\mathrm{T}}(p) W_j(p) \mathrm{d}p = 0, \quad \forall j \neq k, \tag{2.18}$$

which means that vector eigenfunctions corresponding to distinct eigenvalues are *orthogonal with respect to the mass distribution, i.e.*, the mass density distribution $m(p)$ is the weight function of the inner product 2.18. The eigenfunctions corresponding to a particular eigenvalue form a linear manifold [8]. The dimension of this manifold is the finite multiplicity of the eigenvalue. This multiplicity is therefore equal to the largest number of independent eigenfunctions corresponding to that eigenvalue. Those corresponding to an eigenvalue with multiplicity μ are not generally orthogonal to each other, although they are orthogonal to the remaining eigenfunctions and form a linear manifold of dimension μ. By the Gram–Schmidt process, a linear combination of the eigenfunctions can be found, which form an orthogonal set. Any vector function in this set is still a solution of the differential eigenvalue problem 2.11 and 2.12 by the linearity assumption on the differential operator $\mathscr{L}$. By using the degree of freedom on the scale factors, the eigenfunctions can be normalised in order to satisfy the orthonormality condition

$$\int_{\Pi} m(p) W_j^{\mathrm{T}}(p) W_j(p) \mathrm{d}p = 1, \quad \forall j, \tag{2.19}$$

which can also be expressed as

$$\int_{\Pi} m(p) W_j^{\mathrm{T}}(p) W_k(p) \mathrm{d}p = \delta_{jk}, \quad \forall j,k, \tag{2.20}$$

where δ_{jk} denotes the Kronecker symbol. Taking into account that the differential operator $\mathscr{L}$ is self-adjoint and in view of Equation 2.11, also the following result holds:

$$\int_{\Pi} W_j^{\mathrm{T}}(p) \mathscr{L} W_k(p) \mathrm{d}p = \int_{\Pi} W_k^{\mathrm{T}}(p) \mathscr{L} W_j(p) \mathrm{d}p = \lambda_k \delta_{jk}, \quad \forall j,k. \tag{2.21}$$

□

Since the eigenfunctions are orthogonal over the domain Π they constitute a complete set; thus every solution $w(p,t)$ of Equation 2.6, in the case of the free vibration problem, can be expanded in an absolutely and uniform convergent series of the eigenfunctions, W_j, $j = 1,2,\ldots$, in the form

$$w(p,t) = \sum_{j=1}^{\infty} W_j(p) \eta_j(t), \quad p \in \Pi, \tag{2.22}$$

where the scalar function $\eta(t)$ satisfies Equation 2.10 and the initial conditions. This fact is a direct consequence of the so-called *expansion theorem* for the solution of partial differential equations when the method of separation of variables can be applied. The properties of the differential operator $\mathscr{L}$ lead to some interesting properties of the eigenvalue λ. The operator $\mathscr{L}$ is generally positive semi-definite; it is positive definite when it is assumed that rigid displacements are prevented. Moreover, the operator $\mathscr{L}$ is symmetric as a direct consequence of the self-adjointeness property. The property of $\mathscr{L}$ to be positive semi-definite along with Equations 2.20) and 2.21) has the important implication that all the eigenvalues λ_j, $j = 1,2,\ldots$ are non-negative. The square roots of the eigenvalues, $\omega_{n_j} = \sqrt{\lambda_j}$ are the so-called *natural frequencies* of the structure and, consistently with this definition, the eigenfunctions are called *natural modes of vibration* of the structure. The normalised eigenfunctions are called *normal modes*. The eigenfunctions corresponding to zero eigenvalues represent the rigid-body modes, which occur when the structure is partially or fully unconstrained. Now, consider the boundary value problem expressed by Equations 2.6–2.8, representing the partial differential equation along with boundary and initial conditions, which governs the motion of an undamped flexible structure. Suppose that a closed-form solution to the eigenvalue problem is available, then the solution of the boundary value problem can be obtained using the following steps:

(i) Solve the eigenvalue problem 2.11–2.12 for the eigenvalue $\lambda_j = \omega_{n_j}{}^2$ and the eigenfunctions $W_j(p)$.
(ii) If multiple eigenvalues λ_k occur, by the Gram–Schmidt process find a linear combination of the corresponding eigenfunctions to form an orthogonal set.
(iii) Normalise the eigenfunctions in order to satisfy Equations 2.20 and 2.21 and, by virtue of the expansion theorem, assume the solution of the boundary value problem in the form of Equation 2.22.
(iv) Substitute $w(p,t)$ expressed by Equation 2.22 into Equation 2.6 and initial conditions 2.8, left-multiply the result by W_k^{T} and integrate over the domain Π. Take into account orthonormality relations 2.20 and 2.21 and obtain the infinite set of second-order ordinary differential equations known as *modal equations* in the *normal coordinates* $\eta_k(t)$

$$\ddot{\eta}_k(t) + \omega_{n_k}{}^2 \eta_k(t) = F_k(t), \quad k = 1,2,\ldots, \tag{2.23}$$

where

$$F_k(t) = \int_{\Pi} W_k^T(p) f(p,t) \mathrm{d}x, \quad k = 1,2,\ldots \tag{2.24}$$

are the *modal forces* and the set of initial conditions is given by

$$\eta_k(0) = \int_{\Pi} m(p) W_k^{\mathrm{T}}(p) w_0(p) \mathrm{d}p, \quad \dot{\eta}_k(0) = \int_{\Pi} m(p) W_k^{\mathrm{T}}(p) w_1(p) \mathrm{d}p. \tag{2.25}$$

The approach followed in the steps (i)–(iv) is known as *modal analysis*. Unfortunately, a closed-form solution to the differential eigenvalue problem can only be found for a class of very simple problems representative of simple structures under strong assumptions. Examples of the above problems are the well known Sturm–Liouville and Euler–Bernoulli problems, which are problems involving a domain $\Pi \subseteq \mathbb{R}$. The reader is referred to the specialised literature for problems involving domains $\Pi \subseteq \mathbb{R}^2$ or $\Pi \subseteq \mathbb{R}^3$. The presentation of the latter problems is out of the scope of this chapter and the simple cases of a uniform string and a uniform cantilever beam are presented only to illustrate to the reader the form of the eigensolutions.

2.3.1 Uniform String

Consider a string of length L with uniform mass distribution m over the domain $\Pi = (0, L) \subset \mathbb{R}$ and subject to constant tension T high enough to neglect the gravity force. Moreover, assume that dissipative forces do not occur. Under the above strong assumptions, this problem belongs to the class of Sturm–Liouville problems for which closed-form solution of the differential eigenvalue problem can be found. Supposing that the string is fixed at both ends, the transversal displacement $w(p,t)$, supposed small enough, in the case of free vibration, obeys the partial differential equation

$$\frac{\partial^2 w(p,t)}{\partial p^2} = \frac{m}{T}\frac{\partial^2 w(p,t)}{\partial t^2}, \quad 0 < p < L, \tag{2.26}$$

with the boundary conditions

$$w(0,t) = 0, \quad w(L,t) = 0, \tag{2.27}$$

and initial conditions

$$w(p,0) = w_0(p), \quad \dot{w}(p,0) = w_1(p). \tag{2.28}$$

Equation 2.26 along with Equations 2.27 and 2.28 represents the boundary value problem for the string under consideration. Assuming a solution $w(p,t)$ in the form

$$w(p,t) = W(p)\eta(t), \tag{2.29}$$

with $\eta(t)$ subjected to Equation 2.10, the differential eigenvalue problem is expressed by

$$\frac{\mathrm{d}^2 W(p)}{\mathrm{d}p^2} + \lambda \frac{m}{T} W(p) = 0, \tag{2.30}$$

$$W(0) = 0, \quad W(L) = 0, \tag{2.31}$$

which is a Sturm–Liouville problem. The general solution of Equation 2.30 with $\lambda = \omega^2$ can be written as

$$W(p) = c_1 \sin \omega \sqrt{\frac{m}{T}} p + c_2 \cos \omega \sqrt{\frac{m}{T}} p, \quad 0 < p < L. \tag{2.32}$$

By virtue of boundary conditions it is easy to find

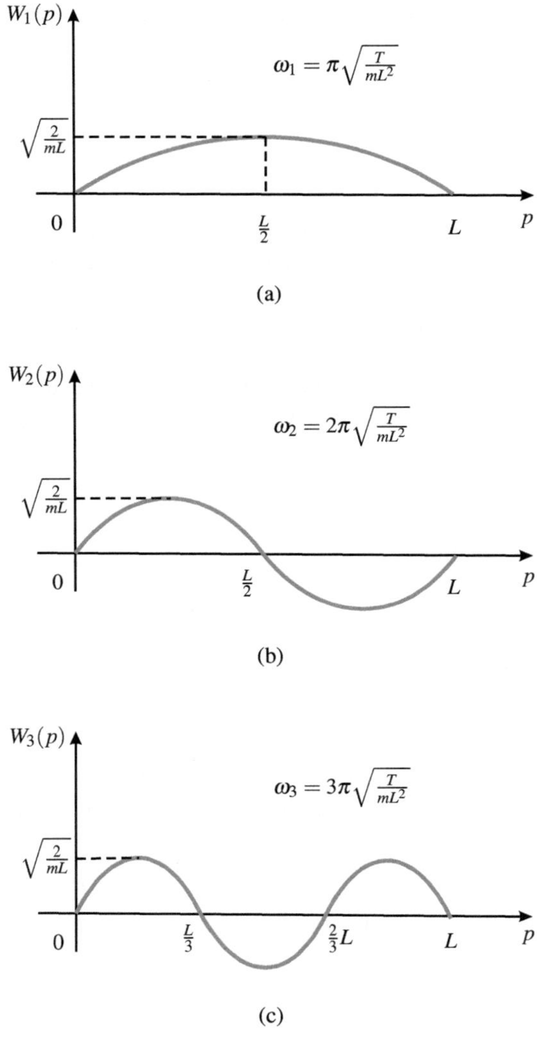

Figure 2.1 (a) First, (b) second and (c) third mode of the uniform string

$$c_2 = 0, \quad \sin\omega\sqrt{\frac{m}{T}}L = 0. \tag{2.33}$$

From the second of relations 2.33 the countable infinite set of eigenvalues is obtained and the natural frequencies are

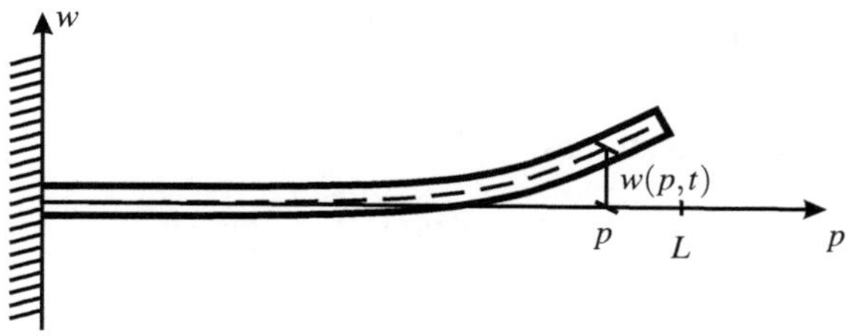

Figure 2.2 Uniform cantilever beam

$$\omega_r = r\pi\sqrt{\frac{T}{mL^2}}, \quad r = 1,2,\dots \tag{2.34}$$

Introducing relation 2.34 into 2.32 the infinite set of orthogonal eigenfunctions is obtained

$$W_r(p) = C_r \sin r\pi\frac{p}{L}, \quad r = 1,2,\dots \tag{2.35}$$

The eigenfunctions can be normalised to satisfy the orthonormality condition

$$\int_0^L mW_r^2(p)\mathrm{d}p = 1, \quad r = 1,2,\dots \tag{2.36}$$

Therefore the infinite set of normal modes is written as

$$W_r(p) = \sqrt{\frac{2}{mL}}\sin r\pi\frac{p}{L}, \quad r = 1,2,\dots \tag{2.37}$$

The shapes of the first three normal modes are depicted in Figure 2.1.

2.3.2 Uniform Cantilever Beam

Consider the bending vibration of the cantilever beam depicted in Figure 2.2. Assume that:

(i) The mass density distribution $m(p)$ is constant over the domain $\Pi = (0,L)$.
(ii) The rotational inertia is negligible.
(iii) The shear deflection is negligible.
(iv) The cross-sectional area moment of inertia is constant so that $I(p) = I$, and thus the bending stiffness EI, where E is the modulus of elasticity, is constant over the domain $(0,L)$.

Under these strong assumptions, by using the extended Hamilton's principle and Lagrangian formalism, the Euler–Bernoulli partial differential equation, which represents the boundary value problem for the cantilever beam under consideration in the free vibration condition, is obtained as

$$-EI\frac{\partial^4 w(p,t)}{\partial p^4} = m\frac{\partial^2 w(p,t)}{\partial t^2}, \quad 0 < p < L, \tag{2.38}$$

subject to the boundary conditions

$$w(0,t) = 0, \quad \left.\frac{\partial w(p,t)}{\partial p}\right|_{p=0} = 0, \tag{2.39}$$

$$\left.\frac{\partial^2 w(p,t)}{\partial p^2}\right|_{p=L} = 0, \quad \left.\frac{\partial^3 w(p,t)}{\partial p^3}\right|_{p=L} = 0, \tag{2.40}$$

and initial conditions

$$w(p,0) = w_0(p), \quad \dot{w}(p,0) = w_1(p). \tag{2.41}$$

Assuming a solution $w(p,t)$ in the form 2.29 with $\eta(t)$ subjected to Equation 2.10, the differential eigenvalue problem 2.11 becomes

$$\frac{\mathrm{d}^4 W(p)}{\mathrm{d}p^4} - \frac{\omega^2 m}{EI}W(p) = 0, \tag{2.42}$$

subject to the boundary conditions, at the fixed end

$$W(0) = 0, \quad \left.\frac{\mathrm{d}W(p)}{\mathrm{d}p}\right|_{p=0} = 0, \tag{2.43}$$

and at the free end

$$\left.\frac{\mathrm{d}^2 W(p)}{\mathrm{d}p^2}\right|_{p=L} = 0, \quad \left.\frac{\mathrm{d}^3 W(p)}{\mathrm{d}p^3}\right|_{p=L} = 0. \tag{2.44}$$

With the notation $\beta^4 = \frac{\omega^2 m}{EI}$, the general solution of the differential eigenvalue problem 2.42 is

$$W(p) = c_1 \sin(\beta p) + c_2 \cos(\beta p) + c_3 \sinh(\beta p) + c_4 \cosh(\beta p). \tag{2.45}$$

Evaluating the boundary conditions, a homogeneous system of algebraic equations is obtained

$$\begin{bmatrix} 0 & 1 & 0 & 1 \\ \beta & 0 & \beta & 0 \\ -\beta^2 \sin(\beta L) & -\beta^2 \cos(\beta L) & \beta^2 \sinh(\beta L) & \beta^2 \cosh(\beta L) \\ -\beta^3 \cos(\beta L) & -\beta^3 \sin(\beta L) & \beta^3 \cosh(\beta L) & \beta^3 \sinh(\beta L) \end{bmatrix} \begin{bmatrix} c_1 \\ c_2 \\ c_3 \\ c_4 \end{bmatrix} = \begin{bmatrix} 0 \\ 0 \\ 0 \\ 0 \end{bmatrix}. \tag{2.46}$$

Equation 2.46 has a nontrivial solution if and only if the matrix on the left-hand side is singular. This happens when

$$\cos(\beta L)\cosh(\beta L) = -1. \tag{2.47}$$

No closed-form expression for the roots of Equation 2.47 is available, thus they should be determined by means of numerical methods. The first three eigenvalues are given approximately by

$$\beta_1 L = 1.875, \quad \beta_2 L = 4.694, \quad \beta_3 L = 7.855, \tag{2.48}$$

and taking into account the definition of the parameter β, the approximate values of the first three natural frequencies are obtained

$$\omega_1 = 1.875^2 \sqrt{\frac{EI}{mL^4}}, \quad \omega_2 = 4.694^2 \sqrt{\frac{EI}{mL^4}}, \quad \omega_3 = 7.855^2 \sqrt{\frac{EI}{mL^4}}. \tag{2.49}$$

The eigenfunctions have the form

$$\begin{aligned} W_r(c) = c_r[&(\sin\beta_r L - \sinh\beta_r L)(\sin\beta_r x - \sinh\beta_r x) + \\ &(\cos\beta_r L + \cosh\beta_r L)(\cos\beta_r x - \cosh\beta_r x)], \quad r = 1, 2, \ldots \end{aligned} \tag{2.50}$$

The constants $c_r, r = 1, 2, \ldots$ can be chosen according to the orthonormalisation procedure. The shapes of the first three modes are depicted in Figure 2.3.

2.4 Approximate Finite-dimensional Models

The example of the uniform cantilever beam illustrated the nature of the boundary conditions. In particular, Equation 2.43 represents geometric boundary conditions and Equation 2.44 natural boundary conditions. Generally, in the formulation of the boundary value problem for a flexible structure two types of boundary conditions are encountered: (a) *geometric boundary conditions* that reflect a geometric condition at the boundary such as zero displacement and/or zero slope, (b) *natural boundary conditions* that reflect force and/or bending moment balance at the boundary. This issue is relevant when looking for approximate methods to produce finite-dimensional models. The first problem that is encountered in modelling complex structures is to formulate the boundary value problem. This is possible for simple structures if some idealisations are made as in the very simple illustrative examples of the uniform clamped string and the uniform cantilever beam. The reader can find the formulation of the boundary value problem for elementary structures like plates, shells, membranes in specialised textbooks, *e.g.*, [3, 5–7].Even if the boundary value problem can be formulated, the solution of the differential eigenvalue problem in a closed-form is not so easy to obtain and even in the simple case of the uniform cantilever beam the eigensolutions are determined using numerical methods. To model complex elastic structures approximate methods are usually envisaged. The motion of continuous elastic structures is modelled, approximately, by a finite series of space-dependent functions that are multiplied by specified time-dependent functions. The former functions are representative of a spatial discretisation process and finite approximate methods differ one from another depending on the spatial discretisation

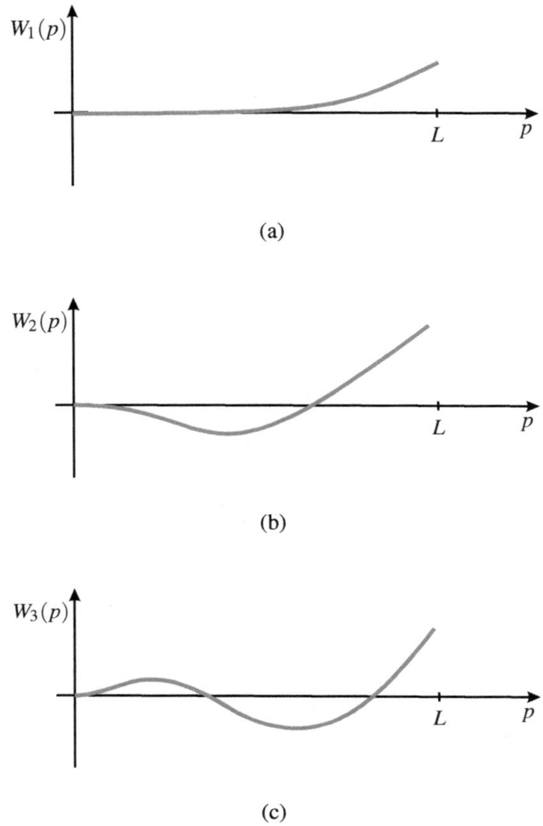

Figure 2.3 (a) First, (b) second and (c) third mode of the uniform cantilever beam

criterion involved, *i.e.*, a global one or a local one. The two most common approximate methods are the *assumed modes* method or Ritz approximation method, which is a global spatial discretisation method, and the *finite element* method, which is an ingenious implementation of the much older Ritz method and is a local one. The selection criteria between the two methods depends on the geometry of the elastic structure. For problems whose geometry is sufficiently simple, the assumed modes method can be implemented since it is possible to select a complete set of space-dependent functions over the global domain Π. By contrast, for problems involving arbitrary geometry, the finite element method is more broadly applicable since the space-dependent functions are local functions defined over small sub domains and identically equal to zero elsewhere, resulting in low-degree polynomials. On the other hand, the finite element method produces discretised models of higher order. The space-dependent functions of interest form three classes of functions depending on the constraints they must satisfy: (i) *eigenfunctions*. They must satisfy the partial differential equation 2.11 and the boundary conditions 2.12; this is a very small class of functions. (ii) *Comparison functions*. They must be $2h$ times differentiable and satisfy all the boundary conditions, *i.e.*, both the geometric and natural ones, but not necessarily the partial differential equation 2.11. Of course, this class will be larger than that of the eigenfunctions. (iii) *Admissible functions*. They must be p times differentiable and are required to satisfy the geometric boundary conditions only, which implies that they constitute a larger class than that of the comparison functions. The space-dependent functions are chosen in order to constitute a complete set of linearly independent functions.

Given a smooth function $f(p)$ defined over the domain Π, consider a set of linearly-independent vector functions $\psi_1(p)$, $\psi_2(p),\ldots,\psi_n(p)$ and define the function $\tilde{f}(p)$ as follows:

$$\tilde{f}(p) = \sum_{r=1}^{n} c_r \psi_r(p), \tag{2.51}$$

where c_r are constant coefficients. If the mean square error

$$\|f-\tilde{f}\|^2 = <f-\tilde{f}, f-\tilde{f}> = \int_{\Pi} (f(p)-\tilde{f}(p))^{\mathrm{T}}(f(p)-\tilde{f}(p))\mathrm{d}p \tag{2.52}$$

can be made less than any arbitrarily small positive number ε by choosing n large enough, then the set $\psi_1(p)$, $\psi_2(p),\ldots,\psi_n(p)$ is said to be a *complete set* [1].

To obtain an approximate finite-dimensional model of the distributed parameter system, select a set of comparison functions ψ_j, $j = 1,2,\ldots,n$ and assume an approximate solution of Equation 2.6 as follows:

$$\tilde{w}(p,t) = \sum_{j=1}^{n} \psi_j(p) q_j(t), \tag{2.53}$$

where n is the number of the generalised coordinates $q_j(t)$, i.e., the number of degrees of freedom of the finite-dimensional system and it is assumed large enough that the set of comparison functions constitutes a complete set in the sense discussed above. To find an approximate solution of the boundary value problem, step (iv) of the procedure described in Section 2.3 can be followed, *i.e.*, substitute Equation 2.53 into Equation 2.6, multiply the result by $\psi_i(p)^{\mathrm{T}}$, $i = 1,2,\ldots,n$ and integrate over the domain Π to obtain

$$\sum_{j=1}^{n} m_{ij}\ddot{q}_j(t) + \sum_{j=1}^{n} k_{ij} q_j(t) = F_i(t), \quad i = 1,2,\ldots,n, \tag{2.54}$$

where

$$m_{ij} = \int_{\Pi} m(p) \psi_i^T(p) \psi_j(p) \mathrm{d}x = m_{ji}, \quad i,j = 1,2,\ldots,n, \tag{2.55}$$

and

$$k_{ij} = \int_{\Pi} \psi_i^T(p) \mathscr{L} \psi_j(p) \mathrm{d}x = k_{ji}, \quad i,j = 1,2,\ldots,n. \tag{2.56}$$

The symmetric property of the coefficients m_{ij} is obvious, that concerning k_{ij} is implied by the fact that the differential operator $\mathscr{L}$ is self-adjoint. The coefficients m_{ij} and k_{ij} represent the mass and stiffness coefficients, respectively, while the functions

$$F_i(t) = \int_{\Pi} \psi_i^T(p) f(p,t) \mathrm{d}x, \quad i = 1,2,\ldots,n \tag{2.57}$$

are the generalised forces.

Equation 2.54 expresses in a compact form a set of n second-order ordinary differential equations, which represents the approximate finite-dimensional model of the flexible structure. Equation 2.54 can be written in the more compact matrix form

$$M\ddot{q}(t) + Kq(t) = F(t), \tag{2.58}$$

where $M = \{m_{ij}\} = M^{\mathrm{T}}$ represents the symmetric positive definite *mass matrix*. The positive definiteness of the matrix M is implied by $m(p) > 0$, $\forall p \in \Pi$. The matrix $K = \{k_{ij}\}$ is the symmetric *stiffness matrix*, it is positive definite if the assumption that $\mathscr{L}$ is positive definite holds, *i.e.*, rigid body motions are prevented. The vector $q(t) = (q_1(t) \quad q_2(t) \quad \ldots \quad q_n(t))^{\mathrm{T}}$ is the generalised coordinates vector and $F(t) = (F_1(t) \quad F_2(t) \quad \ldots \quad F_n(t))^{\mathrm{T}}$ is the generalised force vector. Clearly, given the instantaneous vector of generalised coordinates $q(t)$, the instantaneous deformation of the structure is approximate by the expansion 2.53. Notice that to obtain the solution of Equation 2.58, the initial conditions $q(0)$ and $\dot{q}(0)$ have to be derived first by inverting Equation 2.53 and its time derivative at initial time taking into account the known initial conditions in Equation 2.8.

It should be remarked that the described procedure to derive an approximate finite-dimensional model for the flexible structure under consideration is only formal because it is supposed that a complete set of comparison functions is available in order to obtain a good approximation of the boundary value problem solution, hence it is not easy to implement. However, as clarified in the Introduction of this chapter, the presentation of the boundary value problem, the nature of the eigensolutions and then the approximate finite-dimensional model aims at identifying the structure and at highlighting the properties of the finite-dimensional model expressed by Equation 2.58, which will be exploited both for the implementation of the grey-box identification procedure of Chapter 3 and for the optimal control system design of Chapter 5. The reader is referred to the specialised textbooks cited in the references to deepen the well known approximation methods such as the assumed modes and the finite element models.

2.4.1 Damped Elastic Structures

Equations 2.6–2.8 are representative of the boundary value problem of an undamped elastic structure or an elastic structure with a damping small enough to be neglected. When this assumption is not realistic, the boundary value problem should be reformulated by introducing the contribution of the damping in Equation 2.6. Formally the damping can be introduced by rewriting Equation 2.6 in the form

$$m(p)\ddot{w}(p,t) + \mathscr{D}\dot{w}(p,t) + \mathscr{L}w(p,t) = f(p,t), \quad p \in \Pi, \tag{2.59}$$

where $\mathscr{D}$ denotes the damping operator. The formulation of the boundary value problem is completed by the boundary and initial conditions as in Equations 2.7 and 2.8, respectively. Some assumptions should be made on the nature of the damping operator. The first assumption is that it is a linear self-adjoint operator in order to solve the boundary value problem by using the decoupling procedure as described for the undamped case. Moreover it is assumed that $\mathscr{D}$ is a viscous damping operator known as *proportional damping* operator, resulting from a linear combination of the stiffness operator $\mathscr{L}$ and the mass density m, *i.e.*,

$$\mathscr{D} = \alpha_1 \mathscr{L} + \alpha_2 m \mathscr{I}, \tag{2.60}$$

$\mathscr{I}$ being the identity operator.

This assumption is representative of a special case of great interest. In fact, if the external excitation $f(p,t)$ is harmonic, then the structural damping can be retained to be viscous. It can be proved that the introduction of the damping term does not affect the solution of the differential eigenvalue problem in terms of eigenfunctions or normal mode shapes. By defining the damping coefficient ζ_j as follows:

$$2\zeta_j \omega_{n_j} = \alpha_1 {\omega_{n_j}}^2 + \alpha_2, \quad j = 1,2,\ldots, \tag{2.61}$$

and considering the orthonormality condition that supplements 2.20 and 2.21

$$\int_\Pi W_j^T(p)\mathscr{D}W_k(p)\mathrm{d}x = 2\zeta_j \omega_{n_j} \delta_{jk}, \quad j,k = 1,2,\ldots \tag{2.62}$$

by following the steps (i)–(iv) of the procedure delineated in Section 2.3 for solving the boundary value problem in the undamped case, the infinite set of second-order ordinary differential operators is obtained in terms of normal coordinates as

$$\ddot{\eta}_k(t) + 2\zeta_k \omega_{n_k} \dot{\eta}_k(t) + {\omega_{n_k}}^2 \eta_k(t) = F_k(t), \quad k = 1,2,\ldots \tag{2.63}$$

It should be remarked that, in the case of viscous damping operator $\mathscr{D}$ not satisfying the orthonormality condition 2.62, a closed-form solution of the boundary value problem for damped systems cannot be found.

If an approximate solution of Equation 2.59 subjected to the boundary and initial conditions is pursued, the same procedure described in the case of undamped systems can be followed obtaining the approximate finite-dimensional model

$$M\ddot{q}(t) + D\dot{q}(t) + Kq(t) = F(t), \tag{2.64}$$

where the matrices M and K and the function $F(t)$ have the same meaning as in Equation 2.58) and the entries d_{ij} of the damping matrix D are defined as

$$d_{ij} = \int_{\Pi} \psi_i^{\mathrm{T}}(p)\mathscr{D}\psi_j(p)\mathrm{d}p. \tag{2.65}$$

If $\mathscr{D}$ is the proportional damping operator defined as in 2.60, then the damping matrix D results in a linear combination of the mass and stiffness matrices

$$D = \alpha_1 K + \alpha_2 M. \tag{2.66}$$

This assumption has very important implications. The viscous damping term $D\dot{q}(t)$ with D given by relation 2.66 does not affect the mode shapes that are the eigenvectors of the matrix $M^{-1}K$; the mode shapes are real and orthonormal, as it will be proved later, and have some interesting characteristics. Each mode shape is described by a standing wave that contains fixed stationary node points; the mode shapes diagonalise the matrices M, K and D (see Section 2.4.2). In contrast, if the matrix D is not representative of a proportional damping, the mode shape are complex valued and results in modes, which are described as complex modes, having different characteristics. Each mode shape is described by a travelling wave and appears to contain a moving node point on the structure. The mode shape from the undamped case does not diagonalise the damping matrix D.

2.4.2 Finite-dimensional Systems

Now consider the approximate finite-dimensional model of a flexible structure described in the previous section. Since the introduction of a proportional damping term does not affect the mode shapes from the undamped case all the considerations in this section will be developed for an undamped structure with the aim of highlighting the structure of the state equations, which will be useful for the grey-box identification approach of flexible structures. As usual, define the state vector of the finite-dimensional system as the combination of generalised coordinates and velocities of the finite number of DOFs resulting from the spatial discretisation of the structure, *i.e.*, $z = (q \quad \dot{q})^{\mathrm{T}}$, $q, \in \mathbb{R}^n$, $z \in \mathbb{R}^{2n}$. The system dynamics in 2.58 satisfy the following state space equations:

$$\dot{z}(t) = \begin{pmatrix} 0 & I \\ -M^{-1}K & 0 \end{pmatrix} z(t) + \begin{pmatrix} 0 \\ M^{-1} \end{pmatrix} F(t), \tag{2.67}$$

where $F(t)$ is the vector of generalised forces acting on each DOF, and M and K are the symmetric positive definite mass and stiffness matrices, respectively. The eigensolutions for the system 2.67 are expressed in terms of eigenvalues and eigenvectors of the dynamic matrix

$$A = \begin{pmatrix} 0 & I \\ -M^{-1}K & 0 \end{pmatrix} \in \mathbb{R}^{2n \times 2n}. \tag{2.68}$$

First of all, it is easy to prove the following proposition.

Proposition 2.2. *If the matrices M and K are positive definite, the matrix A in Equation 2.68 has all the eigenvalues on the imaginary axis. The matrix A is diagonalisable and, by similarity transformation, the following matrix $\tilde{A}$ can be obtained:*

$$\tilde{A} = \begin{pmatrix} 0 & I \\ -\Omega & 0 \end{pmatrix}, \tag{2.69}$$

with Ω a diagonal matrix.

Proof. Consider the matrix

$$P = \begin{pmatrix} K & 0 \\ 0 & M \end{pmatrix}. \tag{2.70}$$

Since the matrices K and M are positive definite, the matrix P results to be positive definite. Let $V(z) = z^{\mathrm{T}}Pz$ be a positive definite Lyapunov function candidate. Since

$$A^{\mathrm{T}}P + PA = 0, \tag{2.71}$$

the time derivative of the function $V(z)$ results as $\dot{V}(z) = 0\ \forall z$, and the origin of the state space of the system in Equation 2.67 is stable (see *e.g.*, [4]). Then, by denoting with $\lambda_i(A)$ the ith eigenvalue of A, the matrix A is such that $\mathrm{Re}\,(\lambda_i(A)) \leq 0, \forall i$ (see *e.g.*, [9]). With the same argument on the matrix $-A$, it results $\mathrm{Re}\,(\lambda_i(-A)) \leq 0$, that implies $\mathrm{Re}\,(\lambda_i(A)) = 0, \forall i$, *i.e.*, all the eigenvalues are on the imaginary axis. For an eigenvalue λ_i with algebraic multiplicity $\nu_i \geq 2$ stability implies that $\mathrm{rank}(A - \lambda_i I) = 2n - \nu_i$ (see again [4]). Since $\mathrm{rank}(A - \lambda_i I) = 2n - \nu_i$ if and only if the Jordan block associated with λ_i has order one, A must be diagonalisable, *i.e.*, the matrix A is similar to the matrix

$$\hat{A} = \mathrm{diag}\left\{ \underbrace{j\omega_1, \ldots, j\omega_1}_{\nu_1} \quad \underbrace{-j\omega_1, \ldots, -j\omega_1}_{\nu_1} \quad \cdots \quad \underbrace{j\omega_r, \ldots, j\omega_r}_{\nu_r} \quad \underbrace{-j\omega_r, \ldots, -j\omega_r}_{\nu_r} \right\}, \tag{2.72}$$

supposed to have $2r$ Jordan blocks and $\sum_{i=1}^{r} \nu_i = n$. Since the 2×2 imaginary block $\begin{pmatrix} j\omega_{n_i} & 0 \\ 0 & -j\omega_{n_i} \end{pmatrix}$ can be represented in real form as $\begin{pmatrix} 0 & 1 \\ 0 & -\omega_{n_i}{}^2 \end{pmatrix}$, it is easy to show, by writing the block diagonal matrix and rearranging the indices, that the matrix $\hat{A}$ and hence the matrix A are similar to the matrix $\tilde{A}$ in Equation 2.69 with

$$\Omega = \mathrm{diag}\left\{ \underbrace{-\omega_1^2, \ldots, -\omega_1^2}_{\nu_1} \quad \cdots \quad \underbrace{-\omega_r^2, \ldots, -\omega_r^2}_{\nu_r} \right\}. \tag{2.73}$$

□

This result has very important implications. The eigenvales of matrix $M^{-1}K$ are all positive and, even if the eigenvalue $\lambda_i = \omega_{n_i}{}^2$ has multiplicity ν_i, it is diagonalisable, and ω_{n_i}, $i = 1, \ldots, n$ represent the natural frequencies of the structure. The matrix $M^{-1}K$ has n eigenvectors linearly independent representative of the mode shapes of the structure. The eigenvectors are unique except for a scale factor. By denoting with U the matrix whose columns are the eigenvectors of the matrix $M^{-1}K$, the eigenvalue problem can be written as

$$M^{-1}KU = U\Omega. \tag{2.74}$$

The following proposition can be proved.

Proposition 2.3. *The eigenvector matrix U can be selected to be orthonormal with respect to the mass matrix,* i.e.,

$$U^{\mathrm{T}}MU = I \tag{2.75}$$

and the matrix U diagonalises the stiffness matrix K as

$$U^{\mathrm{T}}KU = \Omega. \tag{2.76}$$

Proof. The eigenvalue problem 2.74, by pre-multiplying the left and right-hand sides of 2.74 by M and, subsequently, by U^{T}, can be rewritten as follows:

$$U^{\mathrm{T}}KU = U^{\mathrm{T}}MU\Omega. \tag{2.77}$$

Equation 2.77 implies that the matrix U, solution of the eigenvalue problem, must satisfy the following equations:

$$U^{\mathrm{T}}MU = \Gamma, \tag{2.78}$$
$$U^{\mathrm{T}}KU = \Gamma\Omega. \tag{2.79}$$

Since M and K are symmetric positive definite matrices, the matrices Γ and $\Gamma\Omega$ are symmetric and positive definite and, moreover, by 2.79 $\Gamma\Omega = \Omega\Gamma$. Since the matrix $\Omega = \mathrm{diag}\omega_{n_i} I_{\nu_i}, i = 1,\ldots,r$, the matrix Γ that satisfies 2.78 and 2.79 has the structure $\Gamma = \mathrm{diag}\Gamma_i, i = 1,\ldots,r$, where Γ_i are symmetric positive definite matrices of order ν_i. The matrix Γ can be factorised as follows:

$$\Gamma = U_\Gamma \Lambda_\Gamma U_\Gamma^{\mathrm{T}}, \tag{2.80}$$

where $U_\Gamma = \mathrm{diag}U_{\Gamma_i}$, $i = 1,\ldots,r$, with U_{Γ_i} orthonormal matrices of order ν_i, and Λ_Γ diagonal. By introducing Equation 2.80 into Equation 2.78 and taking into account the orthonormality property of the matrix U_Γ, Equation 2.78 becomes

$$\tilde{U}^{\mathrm{T}}M\tilde{U} = \Lambda_\Gamma, \quad (\tilde{U} = UU_\Gamma), \tag{2.81}$$

and Equation 2.79 becomes

$$\tilde{U}^{\mathrm{T}}K\tilde{U} = \Lambda_\Gamma U_\Gamma^{\mathrm{T}}\Omega U_\Gamma. \tag{2.82}$$

Taking into account the structure of the matrices U_Γ and Ω and the orthonormality property of the matrices U_{Γ_i}, it is very easy to show that $U_\Gamma^{\mathrm{T}}\Omega U_\Gamma = \Omega$ then Equation 2.82 can be rewritten as

$$\tilde{U}^{\mathrm{T}}K\tilde{U} = \Lambda_\Gamma\Omega. \tag{2.83}$$

By selecting the eigenvector scale factors and renaming $\tilde{U}$ as U the proposition has been proved. □

The columns of the eigenvector matrix U represent the normal mode shapes of the structure. By introducing the transformation in the state space

$$z = \begin{pmatrix} U & 0 \\ 0 & U \end{pmatrix} x, \tag{2.84}$$

where $x = (\eta^{\mathrm{T}} \quad \dot{\eta}^{\mathrm{T}})^{\mathrm{T}}$, η being the so-called *modal coordinates*, and taking into account 2.75 and 2.76, the finite-dimensional model of the undamped structure can be rewritten in terms of modal coordinates as

$$\dot{x}(t) = \begin{pmatrix} 0 & I \\ -\Omega & 0 \end{pmatrix} x(t) + \begin{pmatrix} 0 \\ U^{\mathrm{T}} \end{pmatrix} F(t). \tag{2.85}$$

By considering the finite-dimensional model 2.64 of a damped flexible structure and assuming the damping is a proportional one, from 2.66 it is very easy to verify that the normal mode shapes from the undamped case also diagonalise the damping matrix D; hence the state space model in modal coordinates is

$$\dot{x}(t) = \begin{pmatrix} 0 & I \\ -\Omega & -\Lambda \end{pmatrix} x(t) + \begin{pmatrix} 0 \\ U^{\mathrm{T}} \end{pmatrix} F(t), \tag{2.86}$$

with $\Lambda = \mathrm{diag}\{2\zeta_1\omega_{n_1}, \ldots, 2\zeta_n\omega_{n_n}\}$, $\zeta_l \in (0,1)$ being the damping coefficient of the lth mode.

As will be done both for identification and control purposes in Chapter 3 and Chapters 4 and 5, respectively, assume that the structure is excited by applying concentrated forces in m distinct points of the structure $\bar{p}_1, \ldots, \bar{p}_m$; then the force distribution $f(p,t)$ can be expressed as

$$f(p,t) = \sum_{k=1}^{m} u_k(t)\delta(p - \bar{p}_k), \tag{2.87}$$

where $u_k(t)$ is the force vector applied in the kth point and $\delta(p)$ is the Dirac impulse. Therefore, the generalised forces $F_i(t)$ in 2.57) can be computed as

$$F_i(t) = \int_{\Pi} \psi_i^{\mathrm{T}}(p) \sum_{k=1}^{m} u_k(t)\delta(p - \bar{p}_k)\mathrm{d}p = \sum_{k=1}^{m} \psi_i^{\mathrm{T}}(\bar{p}_k) u_k(t). \tag{2.88}$$

In order to write the equation in a more compact form, define the vector $u(t) = (u_1^{\mathrm{T}}(t) \quad \cdots \quad u_m^{\mathrm{T}}(t))^{\mathrm{T}}$ such that the vector of generalised forces $F(t)$ is

$$F(t) = \Psi^{\mathrm{T}} u(t), \quad \text{with} \quad \Psi = \begin{pmatrix} \psi_1(\bar{p}_1) & \cdots & \psi_n(\bar{p}_1) \\ \vdots & \ddots & \vdots \\ \psi_1(\bar{p}_m) & \cdots & \psi_n(\bar{p}_m) \end{pmatrix}. \tag{2.89}$$

Assuming that the velocity of the structure is observed in the same points where the forces are applied, the output of the system can be computed by differentiating with respect to time Equation 2.53, *i.e.*,

$$y(t) = \begin{pmatrix} \sum_{j=1}^{n} \psi_j(\bar{p}_1)\dot{q}_j(t) \\ \vdots \\ \sum_{j=1}^{n} \psi_j(\bar{p}_m)\dot{q}_j(t) \end{pmatrix} = \Psi\dot{q}(t) = \Psi U\dot{\eta}(t), \tag{2.90}$$

where Equation 2.84 has also been used. Therefore, the state space model in modal coordinates that will be used in the next chapters is

$$\dot{x}(t) = \begin{pmatrix} 0 & I \\ -\Omega & -\Lambda \end{pmatrix} x(t) + \begin{pmatrix} 0 \\ B_2 \end{pmatrix} u(t) = Ax(t) + Bu(t), \tag{2.91}$$

$$y(t) = (0 \quad B_2^{\mathrm{T}})x(t) = B^{\mathrm{T}}x(t), \tag{2.92}$$

where

$$B_2 = U^{\mathrm{T}}\Psi^{\mathrm{T}}. \tag{2.93}$$

In the following, a few properties of this model are recalled.

(i) The system is asymptotically stable, since the eigenvalues of the dynamic matrix are the complex conjugate pairs (λ_l, λ_l^*), with $\lambda_l = -\zeta_l\omega_{n_l} + j\omega_{n_l}\sqrt{1-\zeta_l^2}$, $l = 1, \ldots, n$, and they all have negative real part.

(ii) The system is strictly positive real (see *e.g.*, [4] for more details), since there exist a positive definite matrix Q_0 and a matrix L such that

$$AQ_0 + Q_0A^{\mathrm{T}} = -LL^{\mathrm{T}}, \tag{2.94}$$

$$Q_0C^{\mathrm{T}} = B. \tag{2.95}$$

In fact, it is easy to verify that

$$Q_0 = \begin{pmatrix} \Omega^{-1} & 0 \\ 0 & I \end{pmatrix}, \quad L = \begin{pmatrix} 0 \\ \sqrt{2}\Lambda^{1/2} \end{pmatrix}.$$

(iii) Owing to the above property, the system is minimum-phase.

(iv) Controllability and observability of the modes are equivalent, and a mode is uncontrollable if and only if the corresponding row of the matrix B_2 is zero.

(v) The frequency response matrix is symmetric and has the special form

$$G(j\omega) = B_2^{\mathrm{T}}\Phi(j\omega)B_2, \tag{2.96}$$

with

$$\Phi(j\omega) = \mathrm{diag}\{\phi_1(j\omega), \ldots, \phi_n(j\omega)\}$$

being

$$\phi_l(j\omega) = \frac{j\omega}{\omega_{n_l}^2 - \omega^2 + j2\zeta_l\omega_{n_l}\omega}, \quad l = 1, \ldots, n,$$

and thus the (h,k)th element of $G(j\omega)$ is

$$g_{hk}(j\omega) = \sum_{l=1}^{n} \phi_l(j\omega) b_{hl} b_{kl}, \tag{2.97}$$

where b_{hl} is the (h,l)th element of the input matrix B_2. The symmetry of the frequency response function (FRF) matrix is also known as the *reciprocity property*.

2.5 Conclusions

The chapter presented the basics of the modelling of flexible structures. It started from stating the boundary value problem and presented the theoretical solution based on modal analysis, both in the undamped and damped case. The chapter ended with the most commonly used finite-dimensional approximations, leading to the state space representation in modal coordinates, whose main properties have been recalled. This particular dynamic model will be used in Chapter 3 to tackle the problem of experimental identification *via* a grey-box approach and in Chapters 4 and 5 to carry out the controller design. In particular, the reciprocity property will be exploited for the identification procedure and, along with the diagonal structure of the matrices Ω and Λ, to compute the closed-form solutions of the $\mathscr{H}_2$ and $\mathscr{H}_\infty$ control problems in the colocated control case.

References

1. Arfken G (1985) Mathematical Methods for Physicists, 3rd Ed., Academic Press, Orlando, FL
2. Eringen AC, Suhubi ES (1975) Elastodynamics, vol. II, Linear Theory. Academic Press, New York
3. Gatti PL, Ferrari V (1999) Applied Stuctural and Mechanical Vibrations: Theory, Methods and Measuring Instrumentation, Taylor & Francis, London
4. Khalil H (2002) Nonlinear Systems. 3rd Edition, Prentice-Hall, Upper Saddle River, NJ
5. Meirovitch L (1980) Computational Methods in Structural Dynamics. Sijthoff & Noordhoff, The Nederlands
6. Meirovitch L (1986) Elements of Vibration Analysis. 2nd Ed., McGraw-Hill, New York
7. Meirovitch L (1990) Dynamics and Control of Structures. Wiley-Interscience, New York
8. Stakgold I (2000) Boundary Value Problems of Mathematical Physics, vol. I, SIAM: Classics in Applied Mathematics, New York
9. Zhou K (1998) Essentials of Robust Control. Prentice Hall, Upper Saddle River, NJ

Chapter 3
Identification of Flexible Structures

3.1 Introduction

In the previous chapter many reasons to justify the need for a dynamic model able to describe the behaviour of a flexible structure have been recalled, *e.g.*, simulation or control design purposes. Whatever model one might choose, it cannot be usefully exploited for simulation or control purposes if its parameters, or at least an estimation of them, are not known. Therefore, a model identification technique is necessary. Even though general system identification techniques have been developed in the past years, the parameters of a dynamic model of a flexible structure cannot be easily estimated mainly due to the high-order model that is usually required, and thus the application of a targeted identification procedure is necessary.

A large number of parametric identification techniques for linear systems are available in the literature and a comprehensive list of classical references can be found in [21]. Every parametric identification technique usually starts from a data set that can be both in the time domain and in the frequency domain. The latter has a twofold advantage when the system to be identified is a complex lightly damped flexible structure. First, FRF are usually already available from the structural testing phase. Moreover, it is well known that the effects of measurement noise can be more easily counteracted in the frequency domain. In fact, the FRF can be estimated by using a large number of time domain measurements, even coming from different experiments, so that from a statistical point of view noise can be significantly reduced. For a more thorough discussion on this topic the reader is referred to [34]. Furthermore, with reference to the special model structure based on modal coordinates, some of the parameters have a physical meaning directly related to the frequency domain, *i.e.*, resonant frequencies, and the particular shape of the FRF can also help the user to select the appropriate model order, as will be clear later on.

Among the frequency domain identification methods a distinction can be made between iterative and non-iterative methods. The former resort to nonlinear or mixed linear–nonlinear optimisation algorithms to minimise a cost function (a very recent application to civil structures can be found in [18]), and most of the commercial software packages on structural dynamics are based on such techniques (see [2] and references therein). The latter, besides the commercial PolyMAX method [13, 32] for modal parameter estimation, basically refer to the so-called subspace methods that have several advantages. They naturally provide a multi-input multi-output (MIMO) state space model, which is the representation usually required by optimal control algorithms. They are more suitable for high-order systems since the matrix model is less sensitive to parameter variations than a transfer function. They do not need an estimation of the impulse response function, which for lightly damped systems is usually of poor quality with a limited data set. Finally, compared to iterative methods based on the prediction error method, they do not require any canonical form to be imposed on the system to obtain a state space representation. Such a feature can be very important for MIMO systems. In fact, since no single continuous parametrisation covers all possible MIMO linear systems with a fixed McMillan degree; it may be necessary to change parametrisation during the optimisation routine of the prediction error method, leading to possible local minima [20].

Also, different approaches exist in the literature with reference to the issue of the identification error, *i.e.*, a measure of the distance between "true" and "estimated" models. Classical stochastic methods [21] generally use the identification error as a performance index, while worst-case deterministic approaches directly address this topic [35], providing uncertainty bounds on the model error resulting from measured data (*a posteriori* information) affected by given bounded-norm disturbances and plant *a priori* information [14]. This results in the so-called $\mathscr{H}_\infty$ robust identification [12], with the objective to guarantee good results in spite of *a priori* information correctness. Moreover, frequency domain identification results have been combined with time domain identification and parametric approaches [31, 17]. Recently, quite general results in the ℓ_1 [24] and $\mathscr{H}_\infty$ [25] settings have been presented, deriving structural conditions for robust identification.

The method presented in this book is based on the stochastic approach and is the one originally presented in [3]. The identification procedure is a grey-box technique based on a two-stage algorithm. First, a subspace-based approach is adopted to estimate the dynamic matrix of a state space representation, then a weighted least-squares minimisation allows estimating input and output matrices. Both stages are implemented in the frequency domain and do not require any nonlinear optimisation technique since, as every subspace method, they rely on linear least-square solutions. At the same time, differently from classical black-box subspace methods, the procedure retains the physical meaning of the model parameters, which is a desirable characteristic for a flexible structure model and will be very helpful both for model order reduction and in the optimal controller design phase, as shown in Chapter 5. A similar identification approach is also pursued in [7], where the main objective is to obtain a reconstruction of the mode shapes *via* interpolation of their samples. The target of the identification technique proposed in the present book is to obtain a continuous-time model from sampled data, using a discrete-time model for the first stage and a continuous-time model for the second stage. The grey-box approach is preferred to the black-box one for several reasons. First of all, the assigned structure of the model has a number of properties that will be exploited by the control design phase (passivity, dynamic decoupling, bandpass behaviour), especially those leading to the possibility to derive controllers in closed form [4, 5]. Some of these properties, *e.g.*, minimum phase, are ensured by the colocation of sensor–actuator pairs, which also allows the MIMO feedback control system to provide better performances than for non-minimum phase plants [23] when colocated feedback control is meaningful. For applications when colocated control is not suitable, *e.g.*, control of tip deflection of wings and stabilisers, or when colocated measurements are difficult to procure, an alternative identification procedure is also presented, still with a grey-box approach. As will be shown, the procedure may also be used for experimental modal analysis (see, *e.g.*, [27, 6]) of lightly damped structures. Moreover, the identified model is a state space continuous-time system and all the parameters have a clear physical meaning. Nevertheless, the proposed algorithm takes into account the sampled-data nature of the measurements, thus the relationship between discrete-time and continuous-time frequency response functions is suitably exploited. Furthermore, owing to the particular model structure, *a priori* fixed, the user can easily select the model order and a simple order reduction iterative strategy is implemented during the identification procedure itself in a natural way by exploiting the physical meaning of the entries of the input and output matrices.

The identification procedure can be summarised as follows. The first stage, based on the subspace approach proposed in [21], estimates the eigenvalues of the sampled-data system, which are then mapped to the continuous-time leading to an estimate of natural frequencies and damping coefficients of the flexible structure. The second stage consists of estimating the modal residues directly of the continuous-time model, taking into account the relationship between discrete-time and continuous-time FRFs and by resorting to a weighted least-squares algorithm, where the weights are selected depending on the frequency range so as to improve the quality of identification where the input forces act on the structure. The adoption of a frequency-weighted least-squares technique also improves the results [7]. From the knowledge of residues two different techniques are proposed to identify the input and output matrices. The first one addresses the case when colocated measurements are available, leading to a model with properties satisfying physical constraints such as reciprocity and passivity and reducing the model order at the same time. The second one is devised to tackle the identification problem when colocated measurements are not available and is based on the rank deficient matrix approximation.

3.2 Basics of Subspace-based Identification

This section is dedicated to readers who are not familiar with the methodologies for system identification. It briefly summarises the principles of the subspace methods and the basic algorithms solving linear system identification problem. Readers with experience on this topic can omit this section without any risk of not understanding the following material dedicated to the specific problem of the identification of flexible structures. Readers interested in further understanding the topic can find a good starting point in [36] and a comprehensive guide in [20].

3.2.1 Time Domain Approach

Given a state space realisation of a discrete-time LTI system in the form

$$x_{k+1} = Ax_k + Bu_k, \tag{3.1}$$

$$y_k = Cx_k + Du_k, \tag{3.2}$$

with x_k the n-dimensional state vector at the time instant $k \in \mathbb{N}$, A a $n \times n$ real matrix, B a $n \times m$ real matrix, C a $p \times n$ real matrix and D a $p \times m$ real matrix. To identify this system means to obtain an estimate of these matrices starting from a data set of measured input u_k and output y_k sequences.

The basic idea of all subspace identification methods is to determine the state space matrices from the observability and controllability subspaces[1]. To understand this, it can be helpful to recall the observability matrices

$$\mathcal{O}_n = \begin{pmatrix} C \\ CA \\ \vdots \\ CA^{n-1} \end{pmatrix}, \quad \mathcal{O}_{n+1} = \begin{pmatrix} C \\ CA \\ \vdots \\ CA^{n} \end{pmatrix}. \tag{3.3}$$

Of course, if they were known together with the system order n, the output matrix C could be directly obtained by selecting the first p rows of $\mathcal{O}_{n+1}$ and its remaining rows $\mathcal{O}_C$ would allow computing the dynamic matrix A by observing that

$$\mathcal{O}_C = \begin{pmatrix} CA \\ CA^2 \\ \vdots \\ CA^{n} \end{pmatrix} = \mathcal{O}_n A, \tag{3.4}$$

from which A can be obtained as

$$A = \mathcal{O}_n^{\dagger} \mathcal{O}_C, \tag{3.5}$$

where the symbol $X^{\dagger}(X^{\mathrm{T}}X)^{-1}X^{\mathrm{T}}$ denotes the Moore–Penrose pseudo-inverse of the full-rank matrix X, which is well defined if a minimal realisation[2] is sought. In a dual way, the input matrix B may be read directly from the first m columns of the

[1] These concepts can be found in any textbook on linear system theory, *e.g.*, [19, 33, 30].

[2] A state space realisation of a LTI system is said to be *minimal* if there exists no other realisation of lower order. Minimal realisations are always both controllable and observable, *i.e.*, controllability and observability matrices are both full-rank.

controllability matrices, if they are known, namely

$$\mathscr{C}_n = \begin{pmatrix} B\, AB \cdots A^{n-1}B \end{pmatrix}, \quad \mathscr{C}_{n+1} = \begin{pmatrix} B\, AB \cdots A^{n}B \end{pmatrix}. \tag{3.6}$$

Therefore, the main question is how to estimate observability and/or controllability matrices. This is the fundamental problem of the realisation theory, which starting from an input–output description of the system has the objective to construct a minimal state space model. In the time domain, the preferred input–output description is the sequence of the impulse responses matrices H_k of dimension $p \times m$, also-called *Markov parameters*, which are related to the state space matrices as

$$H_k = \begin{cases} 0 & k < 0 \\ D & k = 0 \\ CA^{k-1}B & k > 0 \end{cases}. \tag{3.7}$$

Note that the matrix D coincides with the first sample of the impulse response sequence, while the rest of the elements can be organised in the so-called *Hankel matrix* of dimensions $p(n+1) \times m(n+1)$

$$\mathscr{H} = \begin{pmatrix} H_1 & H_2 & \cdots & H_{n+1} \\ H_2 & H_3 & \cdots & H_{n+2} \\ \vdots & \vdots & \ddots & \vdots \\ H_{n+1} & H_{n+2} & \cdots & H_{2n+1} \end{pmatrix}. \tag{3.8}$$

In view of Equations 3.3, 3.6 and 3.7 it is straightforward to show that

$$\mathscr{H} = \mathscr{O}_{n+1}\mathscr{C}_{n+1}. \tag{3.9}$$

In conclusion, from the impulse response sequence all system matrices can be determined by finding any full-rank factorisation of the Hankel matrix[3] $\mathscr{H} = \hat{\mathscr{O}}_{n+1}\hat{\mathscr{C}}_{n+1}$, being $\hat{\mathscr{O}}_{n+1}$ and $\hat{\mathscr{C}}_{n+1}$ and estimate of $\mathscr{O}_{n+1}$ and $\mathscr{C}_{n+1}$, respectively. Practically all subspace-based system identification methods deal with this factorisation problem and they differ from each other in the way they estimate the observability matrix and also in how it is used to determine the system matrices. Of course, before such a factorisation can take place, every identification technique has to first solve the problem to estimate the impulse response sequence from a suitable input–output data collection, usually corrupted by noise, and in turn the Hankel matrix. The reader interested in this problem can refer to any classical textbook on system identification, *e.g.*, [21] and references therein.

Assuming an estimate $\hat{\mathscr{H}}$ of the Hankel matrix to be available, any subspace identification technique has to tackle a second important problem, again related to errors affecting the estimate of the impulse response. The noise corrupting the input–output data makes the rank of the Hankel matrix of any order always full, whereas its rank should be equal to the system order n. Therefore, a method to "reduce"

[3] The described procedure is the basic realisation algorithm originally proposed in [16].

the rank of the Hankel matrix should be sought. Most subspace techniques resort to truncation of the singular value decomposition (SVD) of the estimated Hankel matrix, *i.e.*,

$$\hat{\mathscr{H}} = U\Sigma V^{\mathrm{T}} = \begin{pmatrix} U_s & U_n \end{pmatrix} \begin{pmatrix} \Sigma_s & 0 \\ 0 & \Sigma_n \end{pmatrix} \begin{pmatrix} V_s^{\mathrm{T}} \\ V_n^{\mathrm{T}} \end{pmatrix} \simeq U_s \Sigma_s V_s^{\mathrm{T}}, \tag{3.10}$$

where the partitioning into the "signal" and "noise" parts, *i.e.*, selection of the dimension of Σ_s is chosen equal to the desired rank of the approximation that is the chosen system order. Such a reduction is somehow "optimal" in the sense that it minimises the Frobenius norm of the difference between the original matrix and its low-rank approximation [11]. Moreover, the analysis of the singular values can be helpful in deciding the system order; in fact, ideally only the first n singular values should be positive and the rest should be zero. In practice, it might happen that a significant step, *i.e.*, a sudden discontinuity, in the singular values sequence is found and thus the system order can easily be selected.

After the rank reduction step, the factorisation of the Hankel matrix is easily found as

$$U_s \Sigma_s V_s^{\mathrm{T}} = \hat{\mathscr{O}}_{n+1} \hat{\mathscr{C}}_{n+1}, \quad \text{with } \hat{\mathscr{O}}_{n+1} = U_s \Sigma_s^{1/2},\ \hat{\mathscr{C}}_{n+1} = \Sigma_s^{1/2} V_s^{\mathrm{T}}. \tag{3.11}$$

As seen before, the system input and output matrices can be directly obtained from the first m columns of $\hat{\mathscr{C}}_{n+1}$ and from the first p rows of $\hat{\mathscr{O}}_{n+1}$, respectively, while the dynamic matrix A is found as in (3.5) *via* a least-squares solution.

The identification procedure outlined so far requires an estimate of the impulse response sequence; however, different subspace methods exist, which estimate the observability matrix directly from input–output data. Again the interested reader is referred to the specific literature cited above.

3.2.2 Frequency Domain Approach

Differently from the procedure described in the previous section, the subspace identification methods in the frequency domain start from the knowledge of the input–output description that is the FRF matrix of the system, *i.e.*, the M samples of the $p \times m$ complex valued matrix

$$G_i = G(e^{j\omega_i}) = C(e^{j\omega_i} I - A)^{-1} B + D, \quad i = 1, \dots, M, \tag{3.12}$$

where ω_i are the normalised frequencies, *i.e.*,

$$\omega_i = 2\pi f_i T_s, \tag{3.13}$$

f_i being the ith spectral line (measured in [Hz]) in the frequency range of interest $[f_1, f_2]$ and T_s the sampling time (measured in [s]).

If the spectral lines are equally spaced, the identification problem can be easily transformed into the same problem solved in the time domain by exploiting the inverse discrete Fourier transform (IDFT) to compute the impulse response samples from the FRF samples, *i.e.*,

$$H_k = \frac{1}{2M}\sum_{i=0}^{2M} G_i e^{\frac{j2\pi ki}{2M}}, \quad k = 0,\ldots,2M-1, \tag{3.14}$$

where, since the system matrices in Equation 3.12 are all real valued, the samples G_i with $i = M+1,\ldots,2M-1$ can be constructed as

$$G_{M+i} = G^*_{M-i}, \quad i = 1,\ldots,M-1, \tag{3.15}$$

where $(\cdot)^*$ denotes complex conjugate transpose.

Nevertheless, in many applications the spectral lines are not necessarily equally spaced, *e.g.*, for lightly damped structures more samples in the vicinity of resonant frequencies are desirable to enhance accuracy in the reproduction of the true FRF. Therefore, an identification algorithm that estimates the system state space realisation directly from the FRF samples is necessary. In the following, the basic ideas of the algorithm originally proposed in [26][4] are outlined, while all the steps of the algorithm are detailed in Section 3.4, where in the first stage of the procedure proposed in this book they are used to identify the dynamic model of a flexible structure.

Consider the discrete Fourier transform (DFT) of the state space Equations 3.1, 3.2

$$e^{j\omega}X(\omega) = AX(\omega) + BU(\omega), \tag{3.16}$$

$$Y(\omega) = CX(\omega) + DU(\omega), \tag{3.17}$$

where $X(\omega)$, $U(\omega)$ and $Y(\omega)$ denote the DFT of the time domain signals x_k, u_k and y_k, respectively.

Now, denote with $X^i(\omega)$ the state vector resulting from the application of an input $U(\omega) = e_i$, with e_i the $(m \times 1)$ unit vector with one in the ith position and zero elsewhere and define the compound state matrix as

$$X^C(\omega) = \begin{pmatrix} X^1(\omega) & X^2(\omega) & \cdots & X^r(\omega) \end{pmatrix} \in \mathbb{C}^{n\times m}. \tag{3.18}$$

With this notation and in view of Equations 3.16 and 3.17 and the definition of the FRF matrix 3.12, an input–output description of the system can be obtained in the form

$$e^{j\omega}X^C(\omega) = AX^C(\omega) + B, \tag{3.19}$$

[4] In this seminal paper two algorithms are presented, *Algorithm 1* solving the identification problem using the IDFT as recalled, and *Algorithm 2* solving the identification problem directly in the frequency domain.

$$G(e^{j\omega}) = CX^C(\omega) + D. \tag{3.20}$$

The recursive application of Equations 3.19 and 3.20 together with the definition of the observability matrix in 3.3 leads to the key relation

$$\begin{pmatrix} G(e^{j\omega}) \\ e^{j\omega}G(e^{j\omega}) \\ \vdots \\ e^{j(q-1)\omega}G(e^{j\omega}) \end{pmatrix} = \mathscr{O}_q X^C(\omega) + \Gamma \begin{pmatrix} I_m \\ e^{j\omega}I_m \\ \vdots \\ e^{j(q-1)\omega}I_m \end{pmatrix}, \tag{3.21}$$

where I_m is the $m \times m$ identity matrix and Γ is the lower triangular Toeplitz matrix of the Markov parameters

$$\Gamma = \begin{pmatrix} D & 0 & \cdots & 0 \\ CB & D & \cdots & 0 \\ \vdots & \vdots & \ddots & \vdots \\ CA^{q-2}B & CA^{q-3}B & \cdots & D \end{pmatrix} \in \mathbb{R}^{pq\times mq}. \tag{3.22}$$

In order to write Equation 3.21 for all the spectral lines in a compact form, the following matrices can be defined:

$$\mathbf{G} = \frac{1}{\sqrt{M}} \begin{pmatrix} G_1 & G_2 & \cdots & G_M \\ e^{j\omega_1}G_1 & e^{j\omega_2}G_2 & \cdots & e^{j\omega_M}G_M \\ \vdots & \vdots & \ddots & \vdots \\ e^{j(q-1)\omega_1}G_1 & e^{j(q-1)\omega_2}G_2 & \cdots & e^{j(q-1)\omega_M}G_M \end{pmatrix} \in \mathbb{C}^{qp\times mM}, \tag{3.23}$$

$$\mathbf{W}_m = \frac{1}{\sqrt{M}} \begin{pmatrix} I_m & I_m & \cdots & I_m \\ e^{j\omega_1}I_m & e^{j\omega_2}I_m & \cdots & e^{j\omega_M}I_m \\ \vdots & \vdots & \ddots & \vdots \\ e^{j(q-1)\omega_1}I_m & e^{j(q-1)\omega_2}I_m & \cdots & e^{j(q-1)\omega_M}I_m \end{pmatrix} \in \mathbb{C}^{qm\times mM}, \tag{3.24}$$

and

$$\mathbf{X}^C = \frac{1}{\sqrt{M}} \left(X^C(\omega_1)\ X^C(\omega_2)\ \cdots\ X^C(\omega_M) \right) \in \mathbb{C}^{n\times mM}, \tag{3.25}$$

so that Equation 3.21 yields

$$\mathbf{G} = \mathscr{O}_q \mathbf{X}^C + \Gamma \mathbf{W}_m, \tag{3.26}$$

which can be converted into a relation involving only real valued matrices by defining the following matrices:

$$\mathscr{G} = (\mathrm{Re}\,\mathbf{G} \quad \mathrm{Im}\,\mathbf{G}), \tag{3.27}$$

$$\mathscr{W} = (\mathrm{Re}\,\mathbf{W} \quad \mathrm{Im}\,\mathbf{W}), \tag{3.28}$$

$$\mathscr{X} = (\mathrm{Re}\,\mathbf{X}^C \quad \mathrm{Im}\,\mathbf{X}^C), \tag{3.29}$$

as

$$\mathscr{G} = \mathscr{O}_q \mathscr{X} + \Gamma \mathscr{W}. \tag{3.30}$$

Recalling from the previous section that an estimation of the range space of the observability matrix is the first step to identify the A and C matrices, the first step to take towards this objective is to cancel the term $\Gamma \mathscr{W}$. This can be easily done by right-multiplying both sides of (3.30) by the projector onto the null space of $\mathscr{W}$, *i.e.*,

$$\mathscr{W}^{\perp} = I - \mathscr{W}^{\mathrm{T}}(\mathscr{W}\mathscr{W}^{\mathrm{T}})^{-1}\mathscr{W}, \tag{3.31}$$

thus obtaining the new relation

$$\mathscr{G}\mathscr{W}^{\perp} = \mathscr{O}_q \mathscr{X} \mathscr{W}^{\perp}. \tag{3.32}$$

A deep analysis of this relation is carried out in [26] and it is not reported for brevity; the main conclusion is that if a minimal realisation is sought (so that the pair (A,B) is controllable) and if the number of spectral lines M is high enough compared to the system order, then the range spaces of $\mathscr{G}$ and $\mathscr{O}_q$ are equal. In conclusion, if the above hypotheses on the system and on the data are satisfied, an estimation of the observability matrix can be obtained by a singular value decomposition of $\mathscr{G}\mathscr{W}^{\perp}$ in a way similar to what has been done in Equation 3.11, *i.e.*,

$$\mathscr{G}\mathscr{W}^{\perp} = (U_s \quad U_n)\begin{pmatrix} \Sigma_s & 0 \\ 0 & \Sigma_n \end{pmatrix}\begin{pmatrix} V_s^{\mathrm{T}} \\ V_n^{\mathrm{T}} \end{pmatrix} \simeq U_s \Sigma_s V_s^{\mathrm{T}}. \tag{3.33}$$

Hence, in view of Equation 3.32

$$\mathscr{O}_q = U_s, \tag{3.34}$$

from which both A and C matrices can be determined. However, to improve computational efficiency of the whole algorithm, which usually involves matrices of large dimensions especially when flexible structures have to be identified, instead of computing the SVD of $\mathscr{G}\mathscr{W}^{\perp}$, it can be shown (again see [26]) that

$$\mathscr{G}\mathscr{W}^{\perp} = R_{22}^{\mathrm{T}} Q_2^{\mathrm{T}}, \tag{3.35}$$

where R_{22} and Q_2 are obtained by the QR-factorisation

$$(\mathscr{W}^{\mathrm{T}} \; \mathscr{G}^{\mathrm{T}}) = (Q_1 \; Q_2)\begin{pmatrix} R_{11} & R_{12} \\ 0 & R_{22} \end{pmatrix}, \tag{3.36}$$

and thus the range space of $\mathscr{G}\mathscr{W}^{\perp}$ can be more easily obtained by computing the SVD of the R_{22}^{T} matrix since Q_2 is an orthogonal matrix by construction.

Once the matrices A and C have been estimated from the observability matrix, the determination of B and D can be carried out *via* a least-squares technique owing to the linearity of the FRF matrix with respect to these matrices, as it is clear from

Equation 3.12. The solution to the least-squares problem is shown to be unique in [26] again if a minimal realisation is sought (thus the pair (A,C) is observable) and if the number of spectral lines is not less than the system order, and it results that

$$\begin{pmatrix} B \\ D \end{pmatrix} = \begin{pmatrix} \operatorname{Re}\hat{\mathscr{X}} \\ \operatorname{Im}\hat{\mathscr{X}} \end{pmatrix}^{\dagger} \begin{pmatrix} \operatorname{Re}\hat{\mathscr{G}} \\ \operatorname{Im}\hat{\mathscr{G}} \end{pmatrix}, \tag{3.37}$$

with

$$\hat{\mathscr{G}} = \begin{pmatrix} G_0 \\ G_1 \\ \vdots \\ G_M \end{pmatrix}, \quad \hat{\mathscr{X}} = \begin{pmatrix} C(e^{j\omega_0}I - A)^{-1} & I_p \\ C(e^{j\omega_1}I - A)^{-1} & I_p \\ \vdots & \vdots \\ C(e^{j\omega_M}I - A)^{-1} & I_p \end{pmatrix}. \tag{3.38}$$

As a conclusion to this section, it should be emphasised that both subspace methods discussed lead to a black-box model. In the next section the identification of a flexible structure will be tackled pursuing a grey-box approach for all the motivations discussed in the Introduction of this chapter.

3.3 Identification Problem for a Flexible Structure

This section addresses the problem of identifying the unknown parameters of a grey-box model of a generic flexible structure based on modal coordinates as detailed in Chapter 2 (see Equations 2.91 and 2.92) and recalled here for the reader's convenience

$$\dot{x}(t) = Ax(t) + Bu(t) = \begin{pmatrix} 0 & I \\ -\Omega & -\Lambda \end{pmatrix} x(t) + \begin{pmatrix} 0 \\ B_2 \end{pmatrix} u(t), \tag{3.39}$$

$$y(t) = B^{\mathrm{T}}x(t) = \begin{pmatrix} 0 & B_2^{\mathrm{T}} \end{pmatrix} x(t), \tag{3.40}$$

where $\Lambda, \Omega \in \mathbb{R}^{n\times n}$, $B_2 \in \mathbb{R}^{n\times m}$ must be estimated, if inputs $u \in \mathbb{R}^m$ and outputs $y \in \mathbb{R}^m$ are colocated. If p outputs are measured ($p > m$ is the typical case of experimental modal analysis), the degrees of freedom can be suitably sorted such that the outputs colocated to the inputs are first selected, and thus the output equation is written as

$$y(t) = \begin{pmatrix} 0 & C_2 \end{pmatrix} x(t), \tag{3.41}$$

with

$$C_2 = \begin{pmatrix} B_2^{\mathrm{T}} \\ \bar{C}_2 \end{pmatrix} \in \mathbb{R}^{p\times n}, \tag{3.42}$$

with $\bar{C}_2 \in \mathbb{R}^{(p-m)\times n}$ to be estimated as well.

If no colocated input–output couples exist, the output equation assumes the form of Equation 3.41, where in Equation 3.42 only $\bar{C}_2$ exists. It is worth noticing that the selected model structure has a null D term, as imposed by the equation of motion 2.58. This allows the model to possess a frequency response function decaying at high frequencies, which can be useful in the control design phase since no additional filtering actions, otherwise required, are necessary, which alter the nature of the plant to be controlled and can have negative effects on the closed loop system. In other works, *e.g.*, in [7, 28, 29], the D term is introduced to take into account the unmodelled high frequency modes, while in the approach presented here they are taken into account by extending the frequency range of the identification, as detailed below.

Before describing in detail the identification procedure, the first problem to be tackled in the dynamic analysis of any flexible structure has to be addressed, *i.e.*, the selection of the frequency range of interest. The selection has to take into account both the frequency range $[f_1, f_2]$ Hz where the spectrum of the inputs to the system is significant and that the actual flexible structure is infinite dimensional. In fact, for these systems the FRF in a given frequency range depends not only on the modes with natural frequencies within the considered range but also on the lower and higher frequency modes. Therefore, to obtain a good model in $[f_1, f_2]$ the FRF has to be measured on a wider frequency range, say $[f_1', f_2']$ with $f_1' < f_1$ and $f_2' > f_2$, hence the matrix $\boldsymbol{\Omega}$ will contain the natural frequencies from f_1' to f_2'. The actual values of f_1' and f_2' of course depend on the modal density of the flexible structure and on the accuracy desired for the dynamic model; a possible choice is to extend the original range of two octaves, but user experience plays a key role in such a choice.

Now, it is here assumed that the FRFs of the flexible structure are measured using a digital acquisition system, ideally, but not necessarily, the same digital system that will be used to implement the active vibration controller. This implies that the relationship between the FRF of the sampled-data system and the FRF of the continuous-time system has to be taken into account. Denote with $G_i \in \mathbb{C}^{p \times m}, \quad i = 1, \ldots, M$, the measured samples of the frequency response matrix $\tilde{G}(e^{j\bar{\omega}_i})$ of the sampled-data system at the M normalised frequencies $\bar{\omega}_i = 2\pi f_i T_s$, f_i being the M frequency samples in the frequency range $[f_1', f_2']$ Hz (not necessarily uniformly spaced) and T_s the sampling time. The choice of M depends on the modal density of the flexible structure and obviously on the width of the frequency range of interest and it is typically a power of 2 for efficient numerical computation of the fast Fourier transform (FFT). Such measurements are related to the FRF matrix $G(j\omega)$ of the continuous-time system to be identified as [1]

$$\tilde{G}(e^{j\omega T_s}) = \frac{1 - e^{-j\omega T_s}}{T_s} \sum_{\nu=-\infty}^{\infty} \frac{1}{j(\omega + \nu\omega_s)} G(j(\omega + \nu\omega_s)), \tag{3.43}$$

where $\omega_s = 2\pi/T_s$.

It is of great importance to stress that the initial order of the model to be identified $(2n)$ is assumed to be *a priori* fixed, *e.g.*, on the basis of a classical indicator used in experimental modal analysis [6], *i.e.*,

$$H(\omega_i) = \sum_{h=1}^{p} \sum_{k=1}^{m} |g_{hk}(j\omega_i)|, \quad i = 1, \ldots, M, \tag{3.44}$$

$g_{hk}(j\omega_i)$ being the (h,k)th element of the matrix G_i, by counting (approximately and overestimating) its number of peaks by simple inspection. However, the final model order will be automatically selected in the final stage of the identification procedure after the model order reduction.

3.4 Identification Procedure

The proposed identification procedure, divided into two stages, is described in an algorithmic fashion so as to be easily implementable in any programming language. In the first stage, a well assessed subspace identification technique based on the measured frequency response functions of the structure is adopted to obtain an estimate of the natural frequencies and the damping coefficients [21], and in turn of the dynamic matrix in Equation 3.39. The second stage consists of a procedure specifically devised for identification of the input and output matrices (B_2 and C_2) of flexible structures, which includes an iterative technique for model order reduction.

3.4.1 First Stage

The first stage is only briefly described since all the details can be found in [21], to which the reader is referred also for the issue of identification based on data affected by measurement noise. First of all, since the FRF is obtained from measurements performed on the sample data version (at sampling time T_s) of the system, the eigenvalues of the system can be directly estimated in the discrete-time domain by following the five steps:

(i) Construct the matrices $\mathbf{G} \in \mathbb{C}^{qp \times mM}$, $\mathbf{W} \in \mathbb{C}^{qm \times mM}$, $\mathscr{G}$ and $\mathscr{W}$ as in Equations 3.23, 3.24, 3.27 and 3.28, respectively, by choosing the number q such that
$2n < q \leq M - 2n$, so as to ensure the matrix $\begin{pmatrix} \mathscr{W} \\ \mathscr{G} \end{pmatrix}$ is lower rectangular and such that the matrix R_{11} in the next step is invertible.

(ii) Calculate the QR-factorisation

$$\begin{pmatrix} \mathscr{W}^{\mathrm{T}} & \mathscr{G}^{\mathrm{T}} \end{pmatrix} = \begin{pmatrix} Q_1 & Q_2 \end{pmatrix} \begin{pmatrix} R_{11} & R_{12} \\ 0 & R_{22} \end{pmatrix}, \tag{3.45}$$

where $Q_1 \in \mathbb{R}^{2mM \times qm}$, $R_{11} \in \mathbb{R}^{qm \times qm}$. Notice that the choice of q is crucial to ensure that $\mathscr{G}\mathscr{W}^{\perp} = R_{22}^{\mathrm{T}} Q_2^{\mathrm{T}}$ (refer to Section 3.2.2 to see that the range space of $\mathscr{G}\mathscr{W}^{\perp}$ equals the range space of the observability matrix of the system to be identified).

(iii) Calculate the SVD

$$R_{22}^{\mathrm{T}} = \hat{U}\hat{\Sigma}\hat{V}^{\mathrm{T}} = \begin{pmatrix} \hat{U}_s & \hat{U}_o \end{pmatrix} \begin{pmatrix} \hat{\Sigma}_s & 0 \\ 0 & \hat{\Sigma}_o \end{pmatrix} \begin{pmatrix} \hat{V}_s^{\mathrm{T}} \\ \hat{V}_o^{\mathrm{T}} \end{pmatrix}. \tag{3.46}$$

(iv) Determine an estimate of the system dynamic matrix

$$\hat{A} = \hat{U}_s^{\dagger} \begin{pmatrix} J_2\hat{U}_s \\ 0_{p\times n} \end{pmatrix}, \text{ where } J_2 = \begin{pmatrix} 0_{(q-1)p\times p} & I_{(q-1)p} \end{pmatrix}. \tag{3.47}$$

(v) Compute the eigenvalues z of the matrix $\hat{A}$ and transform them to the continuous-time *via* the mapping $1/T_s \log(z)$. This concludes the estimation of ζ_i and ω_{n_i}, and thus the computation of the matrices Ω and Λ.

Note that the choice at step (iv) ensures stability of the identified system [22], but different choices, still ensuring stability, can be made as proposed in [21].

3.4.2 Second Stage

The second stage consists in determining an estimate of B_2 and C_2. Since B_2 is the input matrix of a continuous-time model whereas the FRF data are available for the sampled-data version of the system, the relationship 3.43 between the FRF $G(j\omega)$ of the continuous-time system and the one $\tilde{G}(e^{j\omega T_s})$ of the sampled-data system must be taken into account.

Then, let

$$C_2 = \begin{pmatrix} c_1 & c_2 & \dots & c_n \end{pmatrix} \text{ and } B_2 = \begin{pmatrix} b_1^{\mathrm{T}} \\ b_2^{\mathrm{T}} \\ \vdots \\ b_n^{\mathrm{T}} \end{pmatrix}; \tag{3.48}$$

with this notation, in view of Equation 2.97, the frequency response matrix can be written as

$$G(j\omega) = \sum_{l=1}^{n} \phi_l(j\omega) c_l b_l^{\mathrm{T}} = \sum_{l=1}^{n} \phi_l(j\omega) R_l, \tag{3.49}$$

where $R_l = c_l b_l^{\mathrm{T}} \in \mathbb{R}^{p\times m}$ is the residue matrix corresponding to the lth mode. Now, differently from [8–10]where an output error identification method is used, owing to the proposed grey-box approach, based on the knowledge of the measured FRF, it is possible to directly estimate the continuous-time residues by simply substituting Equation 3.49 into Equation 3.43 written for the ith frequency ω_i and where the infinite sum is truncated to the first N terms, *i.e.*,

$$\sum_{l=1}^{n} \sum_{\nu=-N}^{N} \frac{\phi_l(j(\omega_i + \nu\omega_s))}{j(\omega_i + \nu\omega_s)} R_l \simeq \frac{T_s}{1 - e^{-j\omega_i T_s}} G_i, \quad i = 1, \dots, M, \tag{3.50}$$

where the symbol $\simeq$ refers to the simultaneous approximation of magnitude and phase of the complex valued functions.

In order to properly select the number of replicas N, an estimate of the truncation error ε is computed as follows:

$$\varepsilon = \left\| \frac{1-e^{-j\omega T_s}}{T_s} \sum_{\nu=N+1}^{+\infty} \left(\frac{1}{j(\omega-\nu\omega_s)} G(j(\omega-\nu\omega_s)) + \frac{1}{j(\omega+\nu\omega_s)} G(j(\omega+\nu\omega_s)) \right) \right\|. \tag{3.51}$$

In view of the passivity property and by taking into account its particular structure in Equation 3.49 and that $\omega \in 2\pi[f_1', f_2']$ with $2\pi f_2' << \omega_s$, the FRF matrix can be approximated as

$$G(j\omega) \simeq \frac{1}{j\omega}\bar{R}, \quad \omega \gg 2\pi f_2', \tag{3.52}$$

with $\bar{R} \in \mathbb{R}^{p\times m}$ and thus

$$\frac{1}{j(\omega \pm \nu\omega_s)} G(j\omega \pm \nu\omega_s) \simeq -\frac{1}{\nu^2\omega_s^2}\bar{R} \quad \forall \nu \in [N+1, +\infty). \tag{3.53}$$

With this approximation, the truncation error becomes

$$\varepsilon \simeq \left\| -\frac{2}{\omega_s^2} \frac{1-e^{-j\omega T_s}}{T_s} \sum_{\nu=N+1}^{+\infty} \frac{1}{\nu^2}\bar{R} \right\| = \frac{2}{\omega_s^2} \left\| \frac{1-e^{-j\omega T_s}}{\omega T_s} \omega \Psi_1(N+1)\bar{R} \right\|, \tag{3.54}$$

where $\Psi_1(\cdot)$ is the logarithmic derivative of the gamma function, also known as the *digamma function*. The right-hand side of the last equation can be upper bounded by the quantity

$$\frac{4\pi f_2' \Psi_1(N+1)}{\omega_s^2} \|\bar{R}\| \leq \frac{4\pi f_2'}{N\omega_s^2} \|\bar{R}\|, \tag{3.55}$$

which gives a method to choose N, once a relative truncation error $\varepsilon_r = \varepsilon / \|\bar{R}\|$ has been selected by the user.

With the following notations:

$$\mathscr{R} = \begin{pmatrix} R_1 \\ R_2 \\ \vdots \\ R_n \end{pmatrix} \in \mathbb{R}^{pn\times m}, \ \Upsilon = \begin{pmatrix} \frac{T_s}{1-e^{-j\omega_1 T_s}} G_1 \\ \frac{T_s}{1-e^{-j\omega_2 T_s}} G_2 \\ \vdots \\ \frac{T_s}{1-e^{-j\omega_M T_s}} G_M \end{pmatrix} \in \mathbb{C}^{pM\times m}, \tag{3.56}$$

$$\Xi = \begin{pmatrix} \bar{\phi}_1(j\omega_1)I_p & \bar{\phi}_2(j\omega_1)I_p & \dots & \bar{\phi}_n(j\omega_1)I_p \\ \bar{\phi}_1(j\omega_2)I_p & \bar{\phi}_2(j\omega_2)I_p & \dots & \bar{\phi}_n(j\omega_2)I_p \\ \vdots & \vdots & \ddots & \vdots \\ \bar{\phi}_1(j\omega_M)I_p & \bar{\phi}_2(j\omega_M)I_p & \dots & \bar{\phi}_n(j\omega_M)I_p \end{pmatrix} \in \mathbb{C}^{pM\times pn}, \tag{3.57}$$

with $\bar{\phi}_l(j\omega_i) = \sum_{\nu=-N}^{N} \frac{\phi_l(j\omega_i)}{j(\omega_i+\nu\omega_s)}$, $i = 1, \dots, M$, $l = 1, \dots, n$, Equation 3.50 assumes the compact form

$$\Xi\mathscr{R} = \Upsilon. \tag{3.58}$$

It is evident that Equation 3.58 is linear in the unknowns, hence they can be easily estimated by resorting to a least-squares technique. In particular, a weighted least mean square is proposed, since the model has to accurately reconstruct the measured FRF in the frequency range $[f_1, f_2]$ while the data are available on the wider range $[f_1', f_2']$. To improve the accuracy in the first range, a weighting matrix can be defined so as to penalise the data with $\omega_i \notin 2\pi[f_1, f_2]$, *e.g.*,

$$W = \mathrm{blockdiag}(\alpha I_{M_1}, I_{M_2}, \alpha I_{M_3}), \quad 0 < \alpha < 1, \tag{3.59}$$

where I_{M_i} is the $M_i \times M_i$ identity matrix, with M_1, M_2 and M_3 the number of spectral lines in the intervals $[f_1', f_1)$, $[f_1, f_2]$ and $(f_2, f_2']$, respectively (of course it results in $M_1 + M_2 + M_3 = M$). A criterion for optimally selecting the value of the scalar weight α will be given in Chapter 7, and of course even more complicated weighting matrices can be defined.

Since $\mathscr{R}$ is a real matrix and Ξ and Υ are complex valued, Equation 3.58 can be solved in the real field as (see Lemma 3 in [21])

$$\mathscr{R} = X_W^\dagger \begin{pmatrix} \mathrm{Re}\,\Upsilon \\ \mathrm{Im}\,\Upsilon \end{pmatrix}, \text{ where } X = \begin{pmatrix} \mathrm{Re}\,\Xi \\ \mathrm{Im}\,\Xi \end{pmatrix}, \tag{3.60}$$

thus yielding the estimation of all the residue matrices R_l (with $X_W^\dagger$ denoting the left W-weighted pseudo-inverse of X)[5].

Now, if colocated measurements are available and effective, *e.g.*, when a colocated control problem is addressed, the estimation of B_2 can be easily done by considering Equations 3.42 and 3.49 written for the diagonal elements, *i.e.*,

$$g_{kk}(j\omega) = \sum_{l=1}^{n} \phi_l(j\omega) b_l(k)^2,$$

where $b_l(k)$ is the kth element of the lth row of B_2, and thus, in view of Equation 3.49

$$|b_l(k)| = \begin{cases} \sqrt{R_l(k,k)} & \text{if } R_l(k,k) > 0, \\ 0 & \text{if } R_l(k,k) \le 0, \end{cases} \quad \begin{matrix} l = 1, \ldots, n, \\ k = 1, \ldots, m. \end{matrix} \tag{3.61}$$

Remark 3.1. It is worth noting that the case $R_l(k,k) \le 0$ does not fulfil the assigned model structure, thus if $R_l(k,k) \le 0$ the lth mode is assumed to be not controllable from the kth input ($b_l(k) = 0$, by property (iv) in Section 2.4). Moreover, if a whole row of B_2 is null, then the corresponding mode is not controllable from any input and it can thus be removed from the model. This simple observation is a very efficient way for model order reduction. In fact, to remove a mode, the corresponding row in the matrices Ω, Λ, B_2 and the corresponding column in C_2 have to be removed. However, if a mode is eliminated, all the residues must be recomputed since the

[5] The left W-weighted pseudo-inverse of a full-rank matrix X through a weighting matrix W is here defined as $X_W^\dagger = (X^\mathrm{T} W X)^{-1} X^\mathrm{T} W$.

matrix Ξ in (3.57) changes. The reduction procedure terminates when no further mode is removed.

Notice that at this point only the absolute value of the entries of B_2 has been estimated. In fact, the following proposition holds

Proposition 3.1. *If all the natural frequencies are distinct, the matrix B_2 can be estimated except for pre-multiplication by a diagonal unitary matrix $\hat{Z}$.*

Proof. Let B and $\hat{B}$ be two different input matrices corresponding to the same system matrix A (which has been determined in the first stage), the corresponding state variables, say x and $\hat{x}$, respectively, must be related by

$$\hat{x} = \hat{Z}x, \tag{3.62}$$

where $\hat{Z}$ is a basis transformation matrix that must satisfy the condition

$$A\hat{Z} = \hat{Z}A. \tag{3.63}$$

In the new base, the input and colocated output matrices are expressed as

$$\hat{B} = \hat{Z}B, \tag{3.64}$$
$$\hat{B}^{\mathrm{T}} = B^{\mathrm{T}}\hat{Z}^{-1}, \tag{3.65}$$

and thus

$$B^{\mathrm{T}} = B^{\mathrm{T}}\hat{Z}^{\mathrm{T}}\hat{Z}. \tag{3.66}$$

In view of Equations 3.63 and 3.66, only special change of bases are admissible in Equation 3.62. In particular, from Equation 3.63 it follows that $\hat{Z}$ must be a matrix of the form

$$\hat{Z} = \begin{pmatrix} Z_{22} + Z_{12}\Lambda & Z_{12} \\ -Z_{12}\Omega & Z_{22} \end{pmatrix}. \tag{3.67}$$

Then, from Equation 3.64 and in view of the special structure of the input matrix, it follows that

$$\hat{B}_2 = Z_{22}B_2, \tag{3.68}$$
$$B_2^{\mathrm{T}}Z_{12}^{\mathrm{T}} = 0. \tag{3.69}$$

From Equation 3.69 and in view of the complete controllability of the plant (*i.e.*, no null rows are admissible in the matrix B_2) and assuming B_2 as full-column rank, it follows that $Z_{12} = 0$. In conclusion, from Equation 3.67 results $\hat{Z} = \mathrm{blockdiag}\{Z, Z\}$, with $Z = Z_{22}$. Moreover, it now can be shown that Z must be a unitary matrix, namely $Z^{\mathrm{T}}Z = I$. In fact, from Equation 3.66 it follows that

$$B_2^{\mathrm{T}}(I - ZZ^{\mathrm{T}}) = 0, \tag{3.70}$$

and, since B_2 is full-column rank, Z is unitary. To prove that Z is diagonal it is sufficient to note that, in view of Equation 3.63, the following equations

$$Z\Lambda = \Lambda Z, \quad Z\Omega = \Omega Z$$

hold. Hence, since $\Lambda = \mathrm{diag}\{\lambda_1, \ldots, \lambda_n\}$ and $\Omega = \mathrm{diag}\{\omega_{n_1}^2, \ldots, \omega_{n_n}^2\}$ are diagonal matrices with distinct entries,

$$(\lambda_i - \lambda_j) z_{ij} = (\omega_{n_i}^2 - \omega_{n_j}^2) z_{ij} = 0,$$

implying

$$z_{ij} = 0, \quad i \neq j.$$

□

In order to determine the sign of each element of B_2, the off-diagonal elements of the matrix residues are taken into account as follows. For each row of B_2, the sign of the maximum element is chosen positive, then the sign of the remaining elements of the row is established by using the relationship following from Equations 3.49, and 3.42

$$R_l(h,\bar{k}) = B_2(l,h) B_2(l,\bar{k}) \Rightarrow \mathrm{sign}(B_2(l,h)) = \mathrm{sign}(R_l(h,\bar{k})), \tag{3.71}$$

where $l = 1, \ldots, n$, $h = \{1, \ldots, m\} - \{\bar{k}\}$ and $\bar{k} = \arg\max_{k=1,\ldots,m} |b_l(k)|$.

The procedure ends with the estimation of $\bar{C}_2$ still by considering Equation 3.49, namely

$$c_l(h) = \frac{R_l(h,\bar{k})}{b_l(\bar{k})}, \quad h = m+1, \ldots, p, \; l = 1, \ldots, n, \tag{3.72}$$

with $c_l(h)$ the hth element of the lth column of the matrix C_2.

3.4.3 Identification Procedure with Non-colocated Measurements

The second stage presented above assumes that colocated measurements are available; however there are some applications where this is not true. This happens in all cases when on a flexible structure is not possible to install a colocated actuator/sensor pair or when the sensor colocated with the actuator provides a low-quality signal. For example, this is the typical case of experimental modal analysis tests, when a shaker is used as single point actuation and usually the colocated accelerometer provides a signal corrupted by the high-level noise generated by the shaker itself. Therefore, the corresponding signal is excluded from the data used in the identification procedure. In such cases, the output equation of the model to be identified cannot be written in the form of Equation 3.40 but only in the form of Equations 3.41 and 3.42.

Nevertheless, the steps of the second stage of the algorithm are still valid up to Equation 3.60 and provide an estimate of the modal residues. In order to estimate input and output matrices the definition of the modal residues in Equation 3.49 is taken into account and the relation with C_2 and B_2 as defined in Equation 3.48, namely

$$R_l = c_l b_l^{\mathrm{T}}, \quad l = 1, \ldots, n. \tag{3.73}$$

From this equation it is evident that ideally R_l should be matrices of rank 1, but of course matrices estimated through Equation 3.60 can have rank greater than 1 due to both the measurement noise and errors in pole estimation. This problem can be solved by resorting to the optimal rank 1 approximation of each residue matrix, which is known to be the best approximation in the $\|\cdot\|_2$ norm (see Theorem 2.5.3 in [11]), *i.e.*,

$$\tilde{R}_l = \sigma_1 u_1 v_1^{\mathrm{T}}, \tag{3.74}$$

where σ_1 is the largest singular value of R_l and u_1 and v_1 are the corresponding left and right singular vectors, respectively. By simple inspection of Equations 3.73 and 3.74 an estimate of c_l and b_l is given by the choice

$$c_l = \sqrt{\sigma_1} u_1, \tag{3.75}$$

$$b_l = \sqrt{\sigma_1} v_1, \tag{3.76}$$

made to obtain a balanced realisation of the state space model. Note that in this case the automatic order reduction technique presented for the colocated case cannot be applied since it is based on a property of the model that is valid only if Equation 3.40 is available.

3.5 Application to Experimental Modal Analysis

It is worth emphasising that the identification procedure presented in Section 3.4.2 can be conveniently adopted to derive a dynamic model of the flexible structure for applications where a colocated control strategy has been selected by the control designer. In applications where such a control strategy is not suitable, *e.g.*, control of tip deflection of wings and stabilisers, the identification procedure proposed in Section 3.4.3 can be adopted. The same algorithm is also suitable for experimental modal analysis when colocated measurements are difficult to procure or are of low quality. In fact, the objective of experimental modal analysis [6] is to derive natural frequencies and damping coefficients as well as mode shapes. The first parameters are derived in the first stage of the identification procedure, which, in the terminology typical of experimental modal analysis, is a multi degree of freedom (MDOF) method, while the mode shapes are easily determined from the residues. Here it is assumed that the experimental modal testing phase has been carried out using one or two input points (forces are usually applied using shakers and measured using dynamic load cells located at the shaker axis) and many output points (velocities are nowadays efficiently and accurately measured using automated laser vibrometers) one for each degree of freedom of interest. This testing condition corresponds to having an output matrix C_2 with as many rows as the measuring points and as many columns as the number of modelled modes, and the lth column contains the spatial samples of the lth mode shape, whereas, the input matrix B_2 has one or two columns and as many rows as the number of modelled modes. Therefore, the second stage described in Section 3.4.3 can be applied to estimate C_2 and in turn the mode shapes. The assumption made on the testing protocol is not limiting at all. In fact, even if a "roving hammer" testing technique were used (one or two measuring points, usually adopting accelerometers, and many input points using an instrumented hammer), the results of such a measurement campaign could simply be interpreted as the measurement results of the first type of measurement protocol in view of the reciprocity property, which allows exchanging input and output, thus making the FRF matrix symmetric.

Finally, for the convenience of the interested reader, note that the factorisation choice in Equation 3.75 does not correspond to the normalisation with respect to the mass matrix. Nevertheless, the identification method described so far also allows an easy estimation of the so-called *modal constants* from which any normalisation can be chosen to estimate the mode shapes[6]. Recalling that the modal constants are the constant matrices A_l appearing in the following decomposition of the FRF matrix:

$$G(j\omega) = \sum_{l=1}^{n} \phi_l(j\omega)R_l = j\omega \sum_{l=1}^{n} \left(\frac{A_l}{j\omega - \lambda_l} + \frac{A_l^*}{j\omega - \lambda_l^*} \right), \tag{3.77}$$

it is straightforward to see that the modal constants are related to the residues as

$$A_l = \frac{1}{2j\omega_{n_l}\sqrt{1-\zeta_l^2}} R_l. \tag{3.78}$$

Equation 3.78 shows the classical property that modal constants are purely imaginary for flexible structures exhibiting only stationary waves.

The reader familiar with already existing modal analysis methods has certainly recognised that also the proposed procedure separates the estimation of the system poles from the mode shapes. This is a technique commonly adopted also by many commercial software packages devoted to experimental modal analysis. For example, one of the most widely available packages, in the first step of the identification procedure, uses a polyreference least-squares complex frequency domain method, commonly known as *PolyMAX* [32]. However, differently from the first stage of the procedure described in this book, the cited method estimates the system poles by solving a nonlinear least-squares problem by means of a linearisation technique, thus actually solving a suboptimal problem. Moreover, the method usually overspecifies the model order considerably and then separates the true physical modes from the spurious numerical ones by interpreting a so-called stabilisation diagram, which is aimed at solving a typical system identification problem, namely the determination of the model order. Compared to the model order reduction technique proposed in this book, which is based on physical properties of the structure, the use of the stabilisation diagram is usually time consuming and requires a certain skill of

[6] Details can be found in any textbook on experimental modal analysis, *e.g.*, [6, 15].

the user in setting the thresholds on the parameter values. Furthermore, the poles are estimated by using a polynomial approach and not a geometric one, thus suffering from nonlinearities in the unknown parameters. There are different ways to avoid this nonlinearity, *e.g.*, the so-called *time domain multidegree of freedom* method also usually implemented in commercial software packages. From the knowledge of the impulse response samples, first the coefficients of a polynomial whose roots are the system poles are estimated, and then these can be calculated by using any root finding algorithm. However, such techniques are ill-conditioned for high-order systems.

3.6 Illustrative Examples

In order to highlight all the features and show the capabilities of the proposed identification algorithm, the procedure is applied in two case studies referring to two different applications. The first case study is aimed at identifying a model for control purposes in the simulation of an active vibration control system, while the second case study is a typical experimental modal analysis test.

3.6.1 Identification of a Model for Control

The vibrating system to be actively controlled has eight modes in the frequency range $[20,130]$ Hz and three inputs and three outputs where actuators and sensors are assumed to be present for control implementation.

The identification is firstly performed by considering $M = 16$ spectral lines (in the range $[1,200]$ Hz) of the FRFs of the sampled-data system at $T_s = 1/2000$ s in the noise-free case and fixing an initial number of modes to 10. In order to show the importance of taking into account the relationship between the continuous-time and the sampled-data FRFs through the number of replicas in Equation 3.50, firstly N has been set to 0, which means that just the central term of the infinite sum is considered. The estimated pole parameters $\hat{\zeta}_i$, $\hat{f}_{n_i}$ and the estimated input matrix $\hat{B}_2$ are compared to the real ones ζ_i, f_{n_i} and B_2 by reporting the exact values of the pole parameters, the errors $\Delta\zeta_i\% = 100|\zeta_i - \hat{\zeta}_i|/\zeta_i$ and $\Delta f_{n_i}\% = 100|f_{n_i} - \hat{f}_{n_i}|/f_{n_i}$, as well as the true and estimated input matrices B_2 and $\hat{B}_2$, *i.e.*,

$$\begin{array}{c|c}\zeta & f_n\,[\text{Hz}] \\ \hline 0.010 & 20.559 \\ 0.030 & 25.239 \\ 0.020 & 33.215 \\ 0.050 & 54.329 \\ 0.070 & 56.361 \\ 0.002 & 81.515 \\ 0.080 & 99.074 \\ 0.008 & 128.20 \end{array} \qquad \begin{array}{c|c}\Delta\zeta\% & \Delta f_n\% \\ \hline 1.889\cdot 10^{-5} & 2.105\cdot 10^{-8} \\ 1.661\cdot 10^{-6} & 3.794\cdot 10^{-8} \\ 7.674\cdot 10^{-7} & 2.057\cdot 10^{-8} \\ 2.488\cdot 10^{-8} & 4.237\cdot 10^{-10} \\ 5.191\cdot 10^{-8} & 1.082\cdot 10^{-9} \\ 8.809\cdot 10^{-8} & 5.427\cdot 10^{-12} \\ - & - \\ 8.343\cdot 10^{-10} & 3.493\cdot 10^{-13} \end{array} \tag{3.79}$$

$$B_2 = \begin{pmatrix} 0.3339 & 0.1033 & 0.2350 \\ 0.1672 & 0.0133 & 0.1010 \\ 0.0559 & 0.0799 & 0.0731 \\ 0.7103 & 0.4689 & 0.7718 \\ -0.2531 & -0.0814 & -0.2291 \\ -0.4743 & 1.1671 & 0.5717 \\ 0.0026 & -0.0019 & 0.0020 \\ 0.2528 & 0.6279 & -1.7188 \end{pmatrix}, \quad \hat{B}_2 = \begin{pmatrix} 0.3338 & 0.1027 & 0.2344 \\ 0.1673 & 0.0156 & 0.1017 \\ 0.0560 & 0.0801 & 0.0735 \\ 0.7095 & 0.4665 & 0.7689 \\ 0.2556 & 0.0958 & 0.2400 \\ -0.4742 & 1.1670 & 0.5712 \\ - & - & - \\ -0.2517 & -0.6270 & 1.7181 \end{pmatrix}. \tag{3.80}$$

The algorithm, after the automatic model order reduction, provided a model with only seven modes, while the true system has eight modes, since it is not able to identify the seventh mode because the norm of the associated residue is much lower than the norm of the others. This is evidenced by the blank rows in the pole table and input matrix $\hat{B}_2$. Moreover, note that the sign of some rows of the estimated input matrix is opposite to the sign of corresponding rows of the true input matrix, as predicted in Proposition 3.1. In Figure 3.1 the FRF of the true system and the estimated model are reported together with the spectral lines used as identification data. It can be recognised that even with a reduced order system the reconstruction is quite accurate; also the effect of the sampling is apparent on the phase diagram. If a more accurate model is necessary, *i.e.*, also the almost "uncontrollable" mode must be estimated, the number N of replicas in Equation 3.50 must be increased, and thus by selecting $N = 5300$, value computed by using Equation 3.55 with a relative truncation error $\varepsilon_r = 3 \cdot 10^{-9}$, the results are the following:

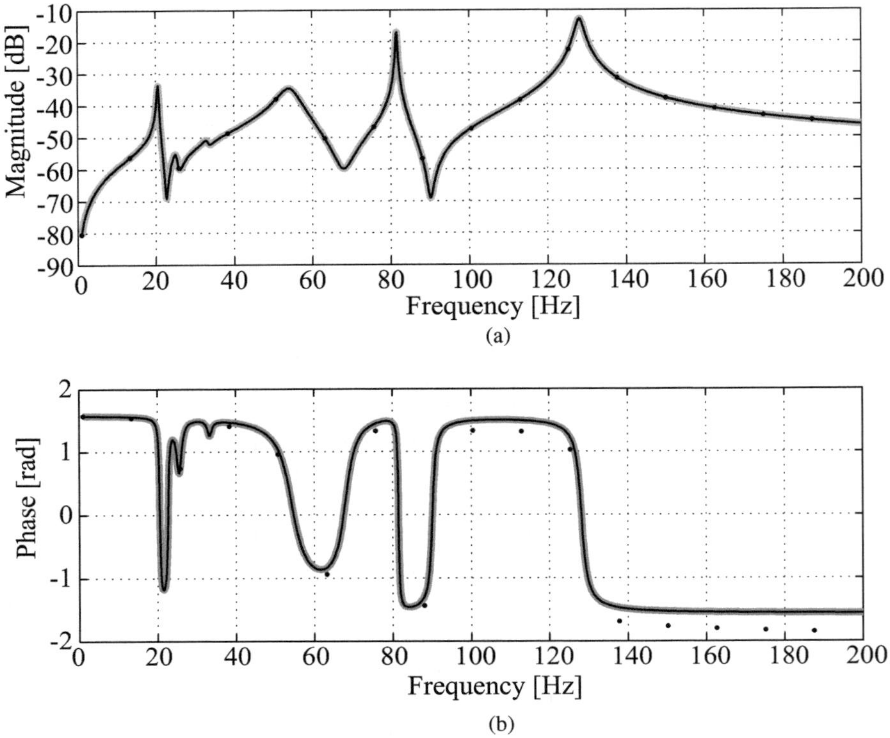

Figure 3.1 (a) Magnitude and (b) phase of the FRF from input 3 to output 3: true system (*solid black lines*), estimated model (*solid grey lines*), spectral lines used by the algorithm (*black dots*)

$\Delta\zeta\%$	$\Delta f_n\%$
$1.889\cdot10^{-5}$	$2.105\cdot10^{-8}$
$1.661\cdot10^{-6}$	$3.794\cdot10^{-8}$
$7.674\cdot10^{-7}$	$2.057\cdot10^{-8}$
$2.488\cdot10^{-8}$	$4.237\cdot10^{-10}$
$5.191\cdot10^{-8}$	$1.082\cdot10^{-9}$
$8.809\cdot10^{-8}$	$5.427\cdot10^{-12}$
$1.349\cdot10^{-5}$	$2.931\cdot10^{-7}$
$8.343\cdot10^{-10}$	$3.493\cdot10^{-13}$

$$\hat{B}_2 = \begin{pmatrix} 0.3339 & 0.1033 & 0.2350 \\ 0.1672 & 0.0133 & 0.1010 \\ 0.0559 & 0.0799 & 0.0731 \\ 0.7103 & 0.4689 & 0.7718 \\ 0.2531 & 0.0814 & 0.2291 \\ -0.4743 & 1.1671 & 0.5717 \\ 0.0026 & -0.0019 & 0.0019 \\ -0.2528 & -0.6279 & 1.7188 \end{pmatrix}. \tag{3.81}$$

It is worth emphasising that the accuracy is not affected by the number of spectral lines used as identification data, since even by increasing it, no improvement is obtained. Only increasing the number of replicas N can improve the accuracy of the identified model parameters. This is a consequence of the absence of measurement noise. Finally, a measurement noise with known statistics is added to the FRFs with the noise model as in [34] and to obtain results with an accuracy comparable to the noise-free case, the number of spectral lines must be increased up to $M = 128$ and

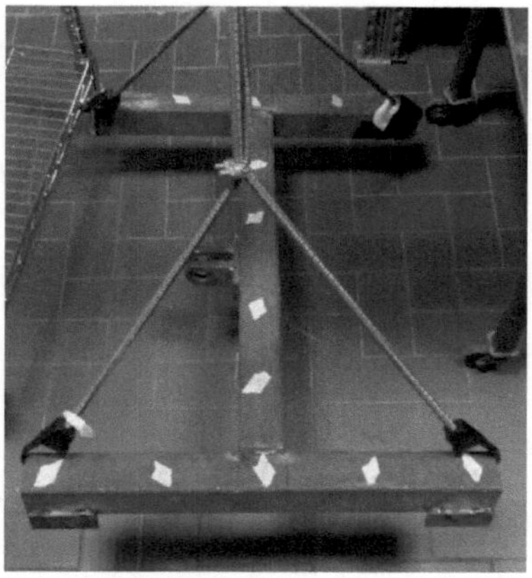

Figure 3.2 Steel structure of the experimental modal analysis test case

the number of replicas up to $N = 8000$. With these choices, and by modifying the first stage according to *Algorithm 2* of [21] to take into account the known noise statistics, the results are

$\Delta\zeta\%$	$\Delta f_n\%$
$8.819 \cdot 10^{-1}$	$1.335 \cdot 10^{-3}$
2.013	$5.301 \cdot 10^{-2}$
1.318	$2.748 \cdot 10^{-3}$
$7.731 \cdot 10^{-1}$	$3.111 \cdot 10^{-3}$
1.578	$1.189 \cdot 10^{-1}$
$7.064 \cdot 10^{-2}$	$7.452 \cdot 10^{-6}$
$1.474 \cdot 10^{-2}$	$3.461 \cdot 10^{-6}$

$$\hat{B}_2 = \begin{pmatrix} 0.3339 & 0.1029 & 0.2350 \\ 0.1683 & 0.0085 & 0.0997 \\ 0.0548 & 0.0784 & 0.0716 \\ 0.6972 & 0.4636 & 0.7605 \\ 0.2871 & 0.1048 & 0.2618 \\ -0.4743 & 1.1671 & 0.5716 \\ -0.2526 & -0.6276 & 1.7186 \end{pmatrix}, \tag{3.82}$$

where it is evident that the algorithm estimates only seven modes. The resulting FRFs are not reported since they do not show any apparent difference with respect to the noise-free case.

3.6.2 Identification for Experimental Modal Analysis

The test article is a "H"-shaped steel structure, with a square section of 0.2×0.2 m positioned in the laboratory using springs to simulate the free–free condition as shown in Figure 3.2. The experimental FRF data base was constructed by using the roving hammer test technique. LMS-TestLab v5.0a Hammer test module version was used to perform the test campaign on the SCADAS III acquisition system. PCB ICPr Hammer model 086C03 (10 mV/lbf) was used together with PCB model 356B07 triaxial ICPr accelerometer (100 mV/g). A complete set of 14 FRFs was

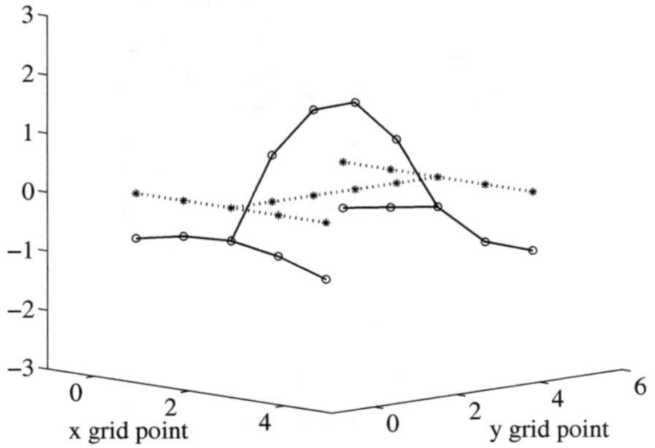

Figure 3.3 Second mode shape (*solid line*) compared to the structure at rest (*dotted line*) with natural frequency 340.38 Hz and damping coefficient 0.0019

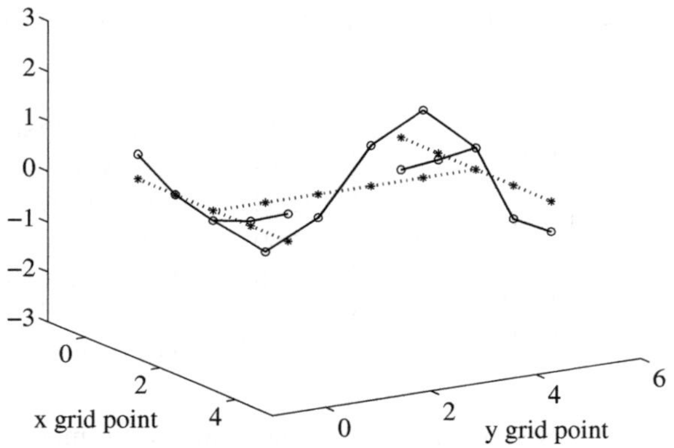

Figure 3.4 Third mode (*solid line*) compared to the structure at rest (*dotted line*) with natural frequency 847.14 Hz and damping coefficient 0.0010

acquired in the frequency range $[1,4096]$ Hz with $M = 4096$ spectral lines, one for each measuring point labelled in the picture.

The identification procedure of Section 3.5 estimated the parameters of 40 modes, *i.e.*, natural frequencies, damping coefficients and mode shapes. For brevity only the parameters of the second and third modes are reported in Figures 3.3 and 3.4, where the mode shapes, plotted in each grid point, are depicted normalised with respect to their maximum amplitude.

3.7 Conclusions

The chapter described a grey-box identification procedure for flexible structures based on the dynamic model derived in Chapter 2. Initially, the chapter recalled some general aspects of the subspace identification technique. Then the identification procedure, divided in two stages, was presented in an algorithmic fashion in order to simplify its implementation in any programming language. The first stage uses a subspace technique, based on the measured frequency response functions, to estimate the natural frequencies and the damping coefficients. The second stage uses a least-squares approach for the identification of the input and output matrices and includes a technique for model order reduction. Furthermore, the chapter showed how the same algorithm can be used in experimental modal analysis. The chapter ends with some illustrative examples of the practical use of the described procedure. This identification procedure will be used to identify the model of the aeronautical panel used for the active vibration control experiments in Chapter 7.

References

1. Åström KJ, Wittenmark B (1991) Computer-Controlled Systems: Theory and Design. 2nd Ed., Prentice Hall, Englewood Cliffs, NJ
2. Balmès E (1996) Frequency Domain Identification of Structural Dynamics Using the Pole/Residue Parametrization. In: Proc. of International Modal Analysis Conference, Dearborn, MI
3. Cavallo A, De Maria G, Natale C, Pirozzi S (2007) Gray-box Identification of Continuous-time Models of Flexible Structures. IEEE Trans Contr Syst Tech, 15:967–981
4. Cavallo A, De Maria G, Natale C, Pirozzi S (2006) Model Identification and Optimal $\mathscr{H}_2$ Vibration Control of an Aeronautical Panel. In: Proc. of ACTIVE 2006, Adelaide
5. Cavallo A, De Maria G, Natale C, Pirozzi S (2006) $\mathscr{H}_\infty$ Strongly Stabilizing Bandpass Controllers for Flexible Systems. In: Proc. of 45th IEEE Conference on Decision and Control, San Diego
6. Ewins DJ (1984) Modal Testing: Theory and Practice. Research Studies Press LTD, Baldock
7. Fleming AJ, Moheimani SOR (2003) Spatial System Identification of a Beam and a Trapezoidal Plate. IEEE Trans on Contr Syst Techn, 11:726–736
8. Gillberg J, Ljung L (2005) Frequency-domain Identification of Continuous-time Output Error Models From Sampled Data. In: Proc. of 16-th IFAC World Congress, Prague
9. Gillberg J, Ljung L (2010) Frequency Domain Identification of Continuous-time Output Error Models, Part I: Uniformly Sampled Data and Frequency Function Approximation. Automatica, 46:1–10
10. Gillberg J, Ljung L (2010) Frequency Domain Identification of Continuous-time Output Error Models, Part II: Non-uniformly Sampled Data and B-spline Output Approximation. Automatica, 46:11–18
11. Golub GH, Van Loan CF (1996) Matrix Computation. 3rd Ed., The Johns Hopkins University Press, Baltimore
12. Gu G, Khargonekar PP (1992) Linear and Nonlinear Algorithms for Identification in $\mathscr{H}_\infty$ with Error Bounds. IEEE Trans Autom Contr, 37:953–963
13. Guillaume P, Verboven P, Vanlanduit S, Van Der Auweraer H, Peeters B (2003) A Polyreference Implementation of the Least-squares Complex Frequency-domain Estimator. In: Proc. of International Modal Analysis Conference, Kissimmee

14. Helmicki AJ, Jacobson CA, Nett CN (1991) Control Oriented System Identification: A Worst-Case/Deterministic Approach in $\mathscr{H}_\infty$. IEEE Trans Autom Contr, 36:1163–1176
15. Heylen W, Lammens S, Sas P (1995) Modal Analysis Theory and Testing. Department of Mecanical Engineering, Katholike Universiteit Leuven, Leuven
16. Ho B, Kalaman R (1966) Efficient Contruction of Linear State Variable Models from Input/Output Functions. Regelungstechnik, 14:545–548
17. Inanc T, Sznaier M, Parrilo P A, Sanchez Pena R S (2001) Robust Identification with Mixed Parametric/Nonparametric Models and Time/Frequency-domain Experiments: Theory and an Application. IEEE Trans Contr Syst Techn, 9:608–617
18. Jin G, Sain MK, Spencer BF Jr (2005) Frequency Domain Systems Identification for Controlled Civil Engineering Structures. IEEE Tran Contr Syst Tech, 13:1055–1062
19. Kailath T (1980) Linear Systems. Prentice Hall, Englewood Cliffs, NJ
20. Katayama T (2005) Subspace Methods for System Identification. Springer, London
21. Ljung L, (1998) System Identification: Theory for The User. 2nd Edition, Prentice-Hall, Upper Saddle River, NJ
22. Maciejowski J M (1995) Guaranteed Stability with Subspace Methods. Syst Contr Lett, 26:153–156
23. Maciejowski J M (1989) Multivariable Feedback Design. Addison-Wesley, Essex, UK
24. Mäkilä PM, Partington JR (1999) On Robustness in System Identification. Automatica, 35:907–916
25. Mäkilä P M, Partington J R (2000) Robustness in $\mathscr{H}_\infty$ Identification. Automatica, 36:1685–1691
26. McKelvey T, Akçay H, Ljung L (1996) Subspace-based Multvariable System Identification from Frequency Response Data. IEEE Trans Autom Contr, 41:960–979.
27. Mendes Maia N M, Montalvao Silva J M (Eds.) (1997) Theoretical and Experimental Modal Analysis, Research Studies Press, Distrib by Wiley
28. Moheimani SOR, Heath WP (2000) Model Correction for a Class of Spatio-temporal Systems. In: Proc of American Control Conference, 3768–3772
29. Moheimani SOR, Heath WP (2002) Model Correction for a Class of Spatio-temporal Systems. Automatica, 38:147–155
30. Ogata K (2004) System Dynamics. Prentice Hall, Englewood Cliffs, NJ
31. Parrilo PA, Sanchez Peña RS, Sznaier M (1999) A Parametric Extension of Mixed Time/Frequency Robust Identification. IEEE Trans Autom Contr, 44:364–369
32. Peeters B, Guillaume P, Van der Auweraer H, Cauberghe B, Verboven P, Leuridan J (2004) Automotive and Aerospace Applications of the PolyMAX Modal Parameter Estimation Method. In: Proc of International Modal Analysis Conference (IMAC 22), Dearborn, MI
33. Rugh WJ (1996) Linear System Theory. Prentice Hall, Englewood Cliffs, NJ
34. Sinha NK, Rao GP (1994) Identification of Continuos-Time Systems. Kluwer, Norwell
35. Smith RS, Doyle JC (1992) Model Validation: A Connection Between Robust Control and Identification. IEEE Trans Autom Contr, 37:942–952
36. Viberg M (1995) Subspace-based Methods for the Identification of Linear Time-invariant Systems. Automatica, 31:1835–1851

Chapter 4
Modal Feedback Control of Flexible Structures

4.1 Introduction

After modelling and identification of the parameters of a flexible structure, the key issue of reducing unwanted vibrations of the system must be addressed.

The vibration reduction problem can generally be formulated as follows. A disturbance source acts on a flexible system, resulting in unwanted vibrations in the structure, and in particular at some specific points of the structure, where the vibration must be minimised. To accomplish such a task, actuators are placed on the structure and a suitable control law is sought. Since the model structure only approximately fits the actual structure and the parameters are identified with approximation errors as in any control problem, the proposed solution must exhibit robustness properties, so as to withstand model and parameter uncertainties.

In the Introduction to this book it has been pointed out that historically the first strategy employed was based on the Lueg's idea of creating a compensating noise wave [18]. Obviously, control strategies have evolved since Lueg's intuition. A more recent version of this approach was presented in [24], where an adaptive version of the procedure was proposed. Following this approach, a number of feedforward adaptive solutions were proposed [11, 10]. However, the fundamental limitation of feedforward strategies is the need for additional transducers and the difficulty in choosing the location of the added sensors. Moreover, as the number of tunable parameters grows, the process of adaption can become cumbersome. Finally, a good model of the secondary path (from controller to error sensor) is needed, and a signal highly correlated to the disturbance must be measured. In [22] different "intelligent" algorithms are compared for this task and applied to the control of a 1-D structure (a flexible beam). Starting from the consideration that any identification procedure reduces to the search for an optimal set of parameters, the analysis compares classic identification algorithms like recursive least-squares to nonlinear optimisation strategies, such as evolutionary genetic algorithms, general regression neural networks, adaptive neuro-fuzzy inference systems. The study shows that genetic algorithms and fuzzy identification perform better on low-frequency resonant modes, while in high frequency modes basing the identification for vibration cancellation on general regression neural networks-based algorithms gives better results. Overall, the opinion of the authors in [22] is that the best results are obtained with genetic algorithms and fuzzy inference systems. However, it has to be stressed that the analysis is based only on a set of simulation studies; thus no final statement can be drawn from the paper, only some suggestions for practitioners.

An interesting approach aimed at reducing noise is active structural acoustic control [11], which is based on the consideration that different structural modes have different sound radiation efficiency, thus contributing differently to noise generation. Designing a control strategy that can selectively reduce mode vibration of the most efficient modes (in terms of noise radiation) can indirectly solve the problem of noise reduction.

Model-free solutions have also been investigated [21], using nonlinear algorithms for estimating amplitudes and phases of disturbances and feedforward strategies. The basic idea for a single tonal disturbance is simply to consider the steady-state component of the effect of the disturbance, which for a SISO system can be reduced to a single complex number (phasor). Also the controller is tuned as a single phasor, thus using only two real parameters and an iterative gradient-based algorithm. The multitonal and the MIMO cases are only extensions of this strategy.

The fundamental limitation that all the above techniques share is that to achieve an attenuation of 20 dB the amplitude error of the cancellation signal must be less than 0.6 dB, and the phase error not larger than 5°.

A possible solution to this drawback is the use of feedback control policies, where the disturbance can be asymptotically rejected without being directly measured, based on the *effect* it produces somewhere in the structure. This is the line of attack followed in this chapter, *i.e.*, controlling flexible systems using feedback strategies. Before discussing feedback strategies in detail, some common characteristics of the problem are pointed out.

Characteristic elements of the vibration problem are the frequency spectrum of the disturbance and the shape of the frequency response of the system (high and low modal density). As far as the spectrum of the disturbance is concerned the extremal cases are white noise, with a spectrum extending over all the frequencies, and tonal disturbance, *i.e.*, a sum of pure sinusoidal waves, with a spectrum comprising only a set of pulses, as in Figure 4.1, where the spectrum of the signal $10(0.7\sin 2\pi 50t + 0.5\sin 2\pi 100t + \sin 2\pi 120t)$ is reported and compared with a (pseudo-)random Gaussian white signal. From the structural point of view, the main difference is between systems with low modal densities, *i.e.*, with few resonant modes within a given frequency bandwidth, as is the case for simple 1-D structures with simple geometric shapes, (*e.g.*, uniform strings) and homogeneity material properties, and, at the other extreme, systems with high modal density, resulting from complex geometries, unevenly distributed materials, coupling of flexural-torsional modes, as in the case of plates with stiffeners. A formal statement of the difference between low and high model density, referring to the description in modal coordinates of the flexible system shown in Equation 2.86, is given by the following definitions:

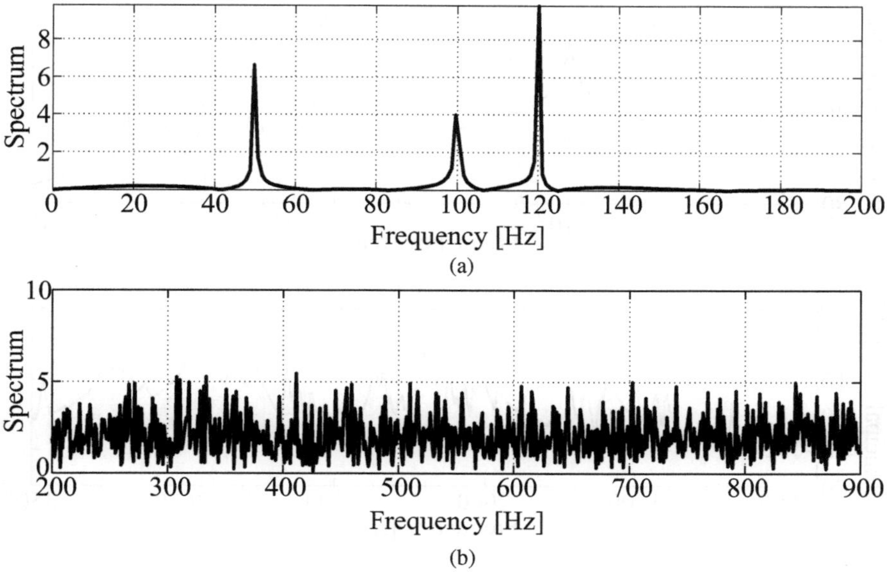

Figure 4.1 Different disturbances: tonal disturbance (a) and broadband disturbance (b)

Definition 4.1. The ith and jth modes of a flexible system are said to be *well separated* if

$$(\omega_{n_i} - \omega_{n_j})^2 \gg (\lambda_i + \lambda_j)(\lambda_i \omega_{n_j} + \lambda_j \omega_{n_i}) \gg 0. \tag{4.1}$$

Note that two modes are well separated if their natural frequencies are far apart and/or they are lightly damped. This definition gives a rigorous formalisation of the "high" and "low" modal density concept. Indeed, systems with low modal density are captured by the following definition.

Definition 4.2. A flexible system is said to have *well separated modes* if all its modes are well separated.

The motivation for the above definitions will be clear in Section 5.10.2 in the next chapter. In Figure 4.2 the largest principal gains of a simple uniform beam and of a complex aeronautic skin panel with stiffeners are compared.

Obviously, the above characteristics can be intermingled, resulting in four different problems:

P1. tonal disturbances with low modal density structure;
P2. tonal disturbances with high modal density structures;
P3. broadband disturbances with low modal density structures;
P4. broadband disturbances with high modal density structures.

Each one of the above problems has its peculiarities, which can be usefully exploited in order to simplify control design. In particular, P1 is usually considered

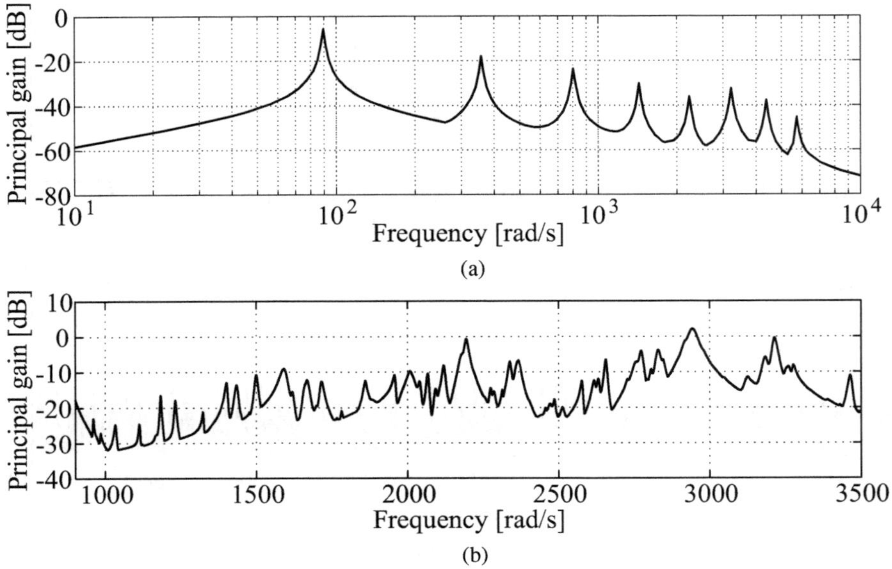

Figure 4.2 Different systems: low modal density (a) and high modal density (b)

a subcase of P2 or P3, depending on the problem. P2 is tackled by using *internal model* control approaches, or adaptive approaches. P3 is solved by *modal control*, with a suitable selection of the modes to work on. Finally, *optimal control* can be used for problem P4 and this will be the topic of Chapter 5. In the sequel the above strategies will be detailed. Before addressing feedback control strategies, a potential problem resulting from feedback must be described, namely the so-called "spillover". This is the subject of the next section.

4.2 Spillover Problem

In principle, the spillover is the effect of unmodelled (usually high-order, *i.e.*, with spectral contents in high frequency) dynamics on a closed loop system employing a controller designed on a finite-dimensional model (and possibly low-order) of the flexible structure [1, 2].

In particular, recall from Chapter 2 that the derivation of a finite-order state model results in an equation of the form of Equation 2.91. It has been shown that if L modes are taken into account, and L is sufficiently high (theoretically infinite), then the finite $2L$-order model is a good approximation of the infinite dimensional system.

Usually, in order to design a controller that can be implemented with standard technology, *i.e.*, a linear time invariant control law, possibly with a reasonably low order, it is customary to consider a suitably low-order model of the flexible system, design a controller for that model and then to validate the design by testing the controller on a high-order model of the plant first, and finally on the true structure.

In order to investigate the effect of this approximation, assume the complete model to be described by using L modes and a *nominal* model computed by considering only N modes, with $L \gg N$. Obviously, the controller must be designed on the $2N$-order model.

Retaining only N modes, a nominal model of order $2N$ is obtained, while the remaining $R = L - N$ modes are called *residual* modes and associated to $2R$-order dynamics. Thus, splitting the state x into a nominal state x_N and a residual state x_R, Equation 2.91 specialises to

$$\dot{x}_N = A_N x_N + B_N u, \tag{4.2}$$

$$\dot{x}_R = A_R x_R + B_R u. \tag{4.3}$$

Also the output is decomposed accordingly, resulting in

$$y = C_N x_N + C_R x_R. \tag{4.4}$$

The term $B_R u$ represents the effect of the input (hence of the controller, when a control law is considered) on the unmodelled, residual dynamics, and is called *control spillover*. Analogously, the term $C_R x_R$ is the contribution of the unmodelled dynamics on the output and is called *observation spillover*.

In order to discuss the effect of spillover on a feedback control system, we refer to a control strategy based on a linear time invariant controller. It has been proved that any stabilising linear controller has the structure of an observer/controller pair [26]; hence the equations of the controller can always be written as

$$\dot{\hat{x}}_N = A_N \hat{x}_N + B_N u + K_o (y - \hat{y}), \tag{4.5}$$

$$\hat{y} = C_N \hat{x}_N, \tag{4.6}$$

for the “state observer”, followed by the “state controller”

$$u = -K_c \hat{x}_N, \tag{4.7}$$

where $\hat{x}_N$ is the controller state, K_o the “observer” gain and K_c the “state controller” gain.

Following the classic approach, we define an observation error variable $e_N = \hat{x}_N - x_N$, whose dynamics is

$$\dot{e}_N = (A_N - K_o C_N)\, e_N + K_o C_R x_R. \tag{4.8}$$

Note that, unlike the classic case, the dynamics of the error is forced by the observation spillover. Moreover, consider the extended state $\xi_N = (x_N^{\mathrm{T}} \quad e_N^{\mathrm{T}})^{\mathrm{T}}$ with dynamics

$$\dot{\xi}_N = \begin{pmatrix} A_N - B_N K_c & B_N K_c \\ 0 & A_N - K_o C_N \end{pmatrix} \xi_N + \begin{pmatrix} 0 \\ K_o C_R \end{pmatrix} x_R. \tag{4.9}$$

The control gains are chosen so as to stabilise this dynamics. In the classic approach the *separation principle* simplifies this task and makes this approach very attractive both from the logical and computational points of view, since it states that it is possible to stabilise separately the matrices $A_N + B_N K_c$ and $A_N - K_o C_N$, under mild hypothesis of stabilisability and detectability of the system. Indeed, since the dynamic matrix in Equation 4.9 is block-triangular, the eigenvalues of the closed loop system are those of the matrices on the diagonal blocks; thus the stability will be retained.

Unfortunately, when considering the residual dynamics the separation principle no longer holds. In fact, defining a state $\zeta = (x_N^{\mathrm{T}} \quad e_N^{\mathrm{T}} \quad x_R^{\mathrm{T}})^{\mathrm{T}}$, one obtains the differential equation

$$\dot{\zeta} = \begin{pmatrix} A_N + B_N K_c & B_N K_c & 0 \\ 0 & A_N - K_o C_N & K_o C_R \\ B_R K_c & B_R K_c & A_R \end{pmatrix} \zeta, \tag{4.10}$$

whose dynamic matrix is no longer triangular and hence stability may be lost.

This is especially true for lightly damped systems where the poles of the residual dynamics may be very close to the imaginary axis. However, Equation 4.10 allows the control designer to also make an interesting observation. Indeed, if there is no control spillover ($B_R = 0$) or observation spillover ($C_R = 0$) the dynamic matrix in Equation 4.10 is again block-triangular; hence closed loop stability is guaranteed by stability of the reduced-order closed loop system and open loop stability of residual dynamics. Thus, it can be concluded that only the *simultaneous* presence of control and observation spillover is harmful for stability.

Based on the above considerations, different strategies for spillover alleviation have been proposed in the literature.

- Since the effect of spillover is harmful for lightly damped modes, a first possible solution is to increase the passive damping by using highly damped materials. The drawback of this approach is the possible increase in weight of the structure.
- Changing the location of sensors and/or actuator by moving them close to the zeros of the mode shapes of the residual modes, in order to avoid excitation and/or sensing of residual modes. However, this solution suffers from different drawbacks, as it is difficult to implement, poorly robust, and, moreover, the control designer seldom has the possibility of deciding the location of sensors and actuator, which is generally selected based on structural considerations.
- Reducing observation spillover by pre-filtering the measurements. The filters must be very selective ("notch" filters) in order to minimise the effect of phase lags, hence this technique is usable only for structures with low modal density and well separated modes. In order to improve robustness with respect to modelling errors and parametric uncertainties phase locked loops (PLL) can be used, which when pre-tuned on estimated frequencies can lock on the actual frequencies.
- Using control action orthogonal to the residual modes. Also this solution may be rather sensitive to modelling errors, if robust control strategies are not used.

- Using symmetric output velocity feedback with colocated sensors and actuators. The idea in this approach can be summarised as follows. Consider for simplicity a high-order description of the modal coordinate of the structure (say, of order $L < \infty$ large enough). Then, from Chapter 2, Equations 2.91 and 2.92 imply that the equation governing the dynamics of the structure can be written as

$$\dot{x} = \begin{pmatrix} 0 & I \\ -\Omega & -\Lambda \end{pmatrix} x + \begin{pmatrix} 0 \\ B_{u2} \end{pmatrix} u, \tag{4.11}$$

$$y = (0 \quad B_{u2}^{\mathrm{T}})x, \tag{4.12}$$

where a velocity output has been assumed. Thus, using a control law of the form

$$u = -(K_1 y + K_2 \dot{y}) \tag{4.13}$$

results in the closed loop system

$$\dot{x} = \begin{pmatrix} 0 & I \\ -(\Omega + B_{u2}K_1B_{u2}^{\mathrm{T}}) & -(\Lambda + B_{u2}K_2B_{u2}^{\mathrm{T}}) \end{pmatrix} x. \tag{4.14}$$

Note that if the control matrices K_1 and K_2 are positive definite, the matrices in Equation 4.14 are symmetric and positive definite, which is enough to guarantee stability of the closed loop system [14], even in the presence of unmodelled dynamics. Of course, the limitation with this approach is that performance is no longer guaranteed in the presence of residual dynamics.

After this brief discussion on the spillover phenomenon, the next section will focus on Problems P1 and P2 presented in Section 4.1, namely the case of systems affected by tonal disturbances.

4.3 Internal Model Control

Internal model control is a classical control strategy [8] that is based on the fundamental property of feedback control: if the model of the plant is very accurate, a good disturbance rejection can be obtained with high gain control (provided that suitable strategies are considered to guarantee the stability of the closed loop system). However, increasing the loop gain has the result of reducing the effect of disturbances of any shape. If the designer is interested in reducing band-limited disturbances it is reasonable to increase the gain only in the particular frequency interval where the disturbance is known to apply. Moreover, if the gain is increased to infinity, a perfect (asymptotic) rejection is guaranteed independently of the magnitude of the disturbance (assuming that the control is not saturated, obviously). For instance, it is well-known that step disturbances can be asymptotically rejected by using a pole at the origin in the compensator. Moreover, the rejection is robust, *i.e.*, it does not depend on the amplitude of the disturbance (as would be the case with a feedforward compensation). Analogously, for disturbances with a band-limited spectrum it is sufficient to increase the open loop gain only in the disturbance band. In particular, tonal disturbances require only tooth comb filters, *i.e.*, undamped filters with frequency resonances at the tones of the disturbance.

In this section first the mathematical basis of the internal model control are outlined, then solutions for adapting the controller gains to an estimate of the disturbance are presented. The basic limitations of the approach are the need for an accurate model of the plant to be controlled, and the applicability to SISO systems only[1].

4.3.1 Basic Approach

One of the first approaches in vibration reduction has been the use of the internal model principle. As mentioned above, the idea is to introduce poles in the control loop that reflect the frequency of tonal disturbances affecting the system, so as to obtain high gains (and high rejecting capabilities of the feedback system) at selected frequencies [25]. Specifically, given an SISO plant with transfer function $G(s)$, assuming a disturbance of the form

$$d(t) = \sum_{i=1}^{\nu} \delta_i \sin(\sigma_i t + \phi_i) \tag{4.15}$$

enters the plant, it can be easily proved that including terms of the form

$$K_1(s) = \prod_{i=1}^{\nu} \frac{1}{s^2/\sigma_i^2 + 1} \tag{4.16}$$

in the control loop guarantees a steady-state rejection of the disturbance. Thus a two-stage procedure is considered. In the first the controller $K_1(s)$ is designed in order to satisfy steady-state requirements and in the second another controller $K_2(s)$ must be designed to stabilise the loop gain $G(s)K_1(s)$ resulting from the first stage. The drawback now is that the transfer function in the second step is more difficult than the original function $G(s)$, due to the increased relative degree of the return ratio, resulting in an increased phase-lag introduced by $K_1(s)$. For this reason in the first stage of the design Equation 4.16 is modified to

$$K_1(s) = \prod_{i=1}^{\nu} \frac{s^2/\rho_i^2 + 2s\zeta_i/\rho_i + 1}{s^2/\sigma_i^2 + 1}, \tag{4.17}$$

[1] Obviously in principle it is possible to apply SISO strategies to MIMO systems with a sequential loop closure, but the result is necessarily a diagonal compensator [19], which is not easy to design, especially if the modes cluster close to each other.

where ρ_i and ζ_i are free design parameters. The choice of these parameters is all but obvious, and this often forces the procedure to rely on a trial-and-error approach, which is not suited to problems with high-order models of the plant to be controlled, as is often the case of flexible systems.

Another drawback of the procedure is that even with tonal problems the frequency of the disturbance is generally not fixed but can vary (slightly) over the time. In order to "follow" the disturbance PLL can be employed, so as to adapt the controller $K_1(s)$ to the actual disturbance. However, when $K_1(s)$ varies $K_2(s)$ has to be redesigned. This explains why sometimes some zeros of $K_1(s)$ are selected with a positive real part, since it is known that zeros with a positive real part "push" the closed loop poles into the negative right half plane. This, however, limit the attainable performances in terms of bandwidth [19].

4.3.2 Adaptive Approach

An alternative approach is the use of adaptive strategies to change the structure of the controller on line. A recent example of this approach was proposed in [17], where the model of the plant is supposed known and the disturbance unknown. It is interesting to outline the procedure followed in [17], in order to show the advantages and the limitations of the approach. The basic idea is to estimate on line the parameters of the disturbance and then design a controller that:

1. has a set of poles varying according to the disturbance estimation and
2. places the closed loop poles at prescribed (fixed) positions in the complex plane.

In order to accomplish such a task the Youla–Kucera parametrisation of all the stabilising controllers of prescribed maximum order is used [26].

In detail, consider the (discrete-time) description of the SISO plant

$$G(z^{-1}) = \frac{z^{-\Delta-1}B(z^{-1})}{A(z^{-1})}, \tag{4.18}$$

where the complex variable z^{-1} represents a unit delay, Δ is a possible plant pure delay measured in number of sampling periods and

$$A(z^{-1}) = 1 + a_1 z^{-1} + \cdots + a_{n_A} z^{-n_A}, \tag{4.19}$$

$$B(z^{-1}) = b_1 + b_2 z^{-2} + \cdots + b_{n_B} z^{-n_B+1}, \tag{4.20}$$

are polynomials in the complex variable z^{-1}. Assuming an output disturbance $d(k)$ affecting the plant additively, as shown in Figure 4.3, the output of the plant is described by[2]

[2] In the sequel, for the sake of notational simplicity, the same symbol z^{-1} is used to denote both the *unit delay* operator in the time domain and the complex variable. The right interpretation depending on the context is left to the reader.

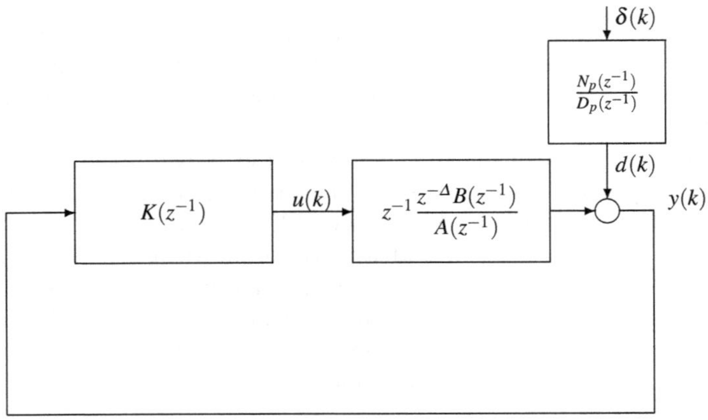

Figure 4.3 Control scheme for internal model control

$$y(k) = \frac{z^{-\Delta-1}B(z^{-1})}{A(z^{-1})}u(k) + d(k). \tag{4.21}$$

Moreover, it is assumed that the disturbance is purely deterministic and hence can be modelled as the impulse response of a linear system with poles on the unit circle

$$d(k) = \frac{N_p(z^{-1})}{D_p(z^{-1})}\delta(k), \tag{4.22}$$

where $\delta(k)$ is the discrete Dirac impulse and the polynomials N_p and D_p are coprime, with degrees[3] n_{N_p} and n_{D_p}, respectively.

The design of the controller $K(z^{-1})$, according to the internal model principle, requires that the controller pole polynomial include D_p, *i.e.*,

$$K(z^{-1}) = \frac{R(z^{-1})}{S(z^{-1})} = \frac{R(z^{-1})}{D_p(z^{-1})S'(z^{-1})}, \tag{4.23}$$

where the zero polynomial R and the pole polynomial S have degrees n_R and n_S, respectively.

According to the Youla–Kucera parametrisation of all the stabilising controllers, R and S can be expressed as

$$R(z^{-1}) = R_0(z^{-1}) + A(z^{-1})Q(z^{-1}), \tag{4.24}$$
$$S(z^{-1}) = S_0(z^{-1}) - z^{-\Delta-1}B(z^{-1})Q(z^{-1}), \tag{4.25}$$

where $Q(z^{-1}) = q_0 + q_1 z^{-1} + \cdots + q_{n_Q} z^{-n_Q}$ is a polynomial whose coefficients are free parameters and R_0 and S_0 are polynomials obtained by solving the Diophantine equation

[3] The degree of a polynomial $W(z^{-1})$ in the indeterminate z^{-1} is the degree of the polynomial $W(\zeta)$ obtained with the substitution $\zeta = z^{-1}$.

$$P(z^{-1}) = A(z^{-1})S_0(z^{-1}) + z^{-d-1}B(z^{-1})R_0(z^{-1}), \tag{4.26}$$

and P is the closed loop pole polynomial, with degree $n_P \leq n_A + n_B + d$. The minimal degree solutions of Equation 4.26 yields $n_{S_0} = N_B + d$ and $n_R = n_A - 1$.

After computation of S_0, a second Diophantine equation is solved,

$$S'(z^{-1})D_p(z^{-1}) + z^{-d-1}B(z^{-1})Q(z^{-1}) = S_0(z^{-1}), \tag{4.27}$$

in the unknowns S' and Q. The latter equation has a unique solution if $n_{S_0} \leq n_{D_p} + n_B + \Delta$ (this is the case, if the minimal degree solution of Equation 4.26 has been selected), $n_{S'} = n_B + d$ and $n_Q = n_{D_p} - 1$.

In conclusion, Equation 4.26 can be solved *a priori*, then, if the model of the disturbance can be estimated online (actually, only the position of the roots of the polynomial D_p are needed, as in the general formulation of the internal model principle), the solution of Equation 4.27 is computed at each sampling time, and the controller updated, thus resulting in an *indirect adaptive control* approach.

As an alternative solution, a *direct adaptive control* strategy can be selected, looking directly for the parameters of the polynomial Q.

In the following, only the algorithm addressing this second approach is reported. The interested reader is referred to [17] for the proof of the stability of the control law. Preliminarily, define the estimate of the polynomial Q at step k as

$$\hat{Q}(z^{-1},k) = \hat{q}_0(k) + \hat{q}_1(k)z^{-1} + \cdots + \hat{q}_{n_Q}(k)z^{-n_Q}, \tag{4.28}$$

and define the parameter vector at step k as $\hat{\theta}(k) = (\hat{q}_0(k) \quad \hat{q}_1(k) \quad \ldots \quad \hat{q}_{n_Q}(k))^{\mathrm{T}}$.

In order to update the estimate of $\hat{\theta}$ at each sample time, the following steps are needed:

1. Compute the value
$$w(k) = A(z^{-1})y(k) - z^{-\Delta-1}B(z^{-1})u(k).$$
2. Compute the value
$$w_1(k) = \frac{S_0(z^{-1})}{P(z^{-1})}w(k).$$
3. Compute the value
$$w_2(k) = \frac{z^{-\Delta-1}B(z^{-1})}{P(z^{-1})}w(k).$$
4. Update the vector
$$\phi(k) = (w_2(k) \quad \ldots \quad w_2(k-n_Q))^{\mathrm{T}}.$$
5. Compute the *a priori* error
$$\varepsilon^0(k) = w_1(k) - \hat{\theta}^{\mathrm{T}}(k-1)\phi(k-1).$$
6. Compute the *a posteriori* error

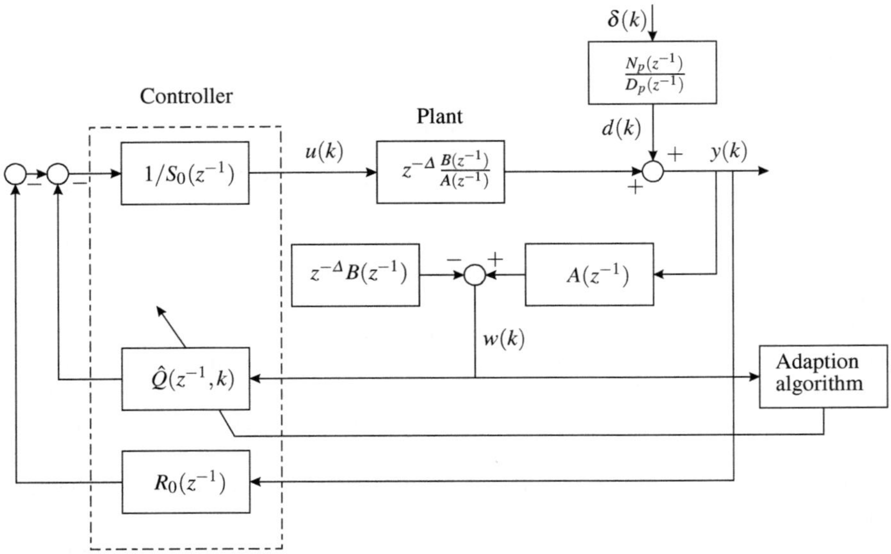

Figure 4.4 Adaptive control scheme

$$\varepsilon(k) = \frac{\varepsilon_0(k)}{1+\phi^{\mathrm{T}}(k-1)F(k-1)\phi(k-1)}.$$

7. Finally, update $\hat{\theta}$ with

$$\hat{\theta}(k) = \hat{\theta}(k-1) + F(k-1)\phi(k-1)\varepsilon(k).$$

and update the gain matrix

$$F(k) = \frac{1}{\lambda_1(k)}\left[F(k-1) - \frac{F(k-1)\phi(k-1)\phi^{\mathrm{T}}(k-1)F(k-1)}{\frac{\lambda_1(k)}{\lambda_2(k)} + \phi^{\mathrm{T}}(k-1)F(k-1)\phi(k-1)}\right],$$

where λ_1 and λ_2 are *forgetting factors*. The simplest way to select them is $\lambda_1(k) = \lambda_2(k) = 1$, which is suitable for systems where the disturbance does not vary with time, since this gives decreasing gains [16]. Otherwise, a variable forgetting factor, with $\lambda_2(k) = 1$ and $\lambda_1(k) = \lambda_0\lambda_1(k-1) + 1 - \lambda_0$, with $0.5 \leq \lambda_0 \leq 0.99$ and $0.95 \leq \lambda_1(0) \leq 0.99$ is proposed.

After the above steps an updated value of $\hat{Q}$ is obtained, and the control law is obtained from

$$S_0(z^{-1})u(k) = -R_0(z^{-1})y(k) - \hat{Q}(z^{-1},k)w(k). \tag{4.29}$$

The scheme of the control loop is shown in Figure 4.4. The effectiveness of the control law has been tested both in simulation and with experimental results for the control of an active suspension, with good results. The direct approach outperforms the indirect one since, as the tonal disturbance is rejected the useful signal-to-noise ratio for identification diminishes, and the identification of the disturbance model parameters becomes less and less effective, leaving a "bias" on the estimated frequency, as noted by the authors [17].

The material presented in this section has addressed the problem of rejecting tonal disturbances, thus covering both problems P1 and P2 of Section 4.1. However, some issues are still left open by this approach, the most important one being the extension to the MIMO case[4] and the increased complexity in the adaptive case. Starting from the next section the focus will be on systems rather than disturbances, and the main difference will be between low-modal and high-modal density, thus addressing problems P3 and P4.

4.4 Modal Control

In this section the problem of counteracting broadband and tonal disturbances for systems with low modal density is considered [5, 12, 23]. As suggested in the Introduction, in this case the best thing to do is to increase damping, especially, in the case of tonal disturbances, for modes with natural frequencies close to that of disturbance. Thus a frequency selected action is sought. For the sake of simplicity, the strategy is presented in two steps: first a state feedback control law is illustrated and its properties analysed. Next the use of a robust observer (loop transfer recovery (LTR) [9], sliding mode [4]) is proposed.

The idea is to consider the description of the system in modal coordinates and to increase the damping of only a prescribed set of modes by using a feedback gain matrix derived from the eigenvectors of the modes to be rejected.

In order to present the strategy, consider the case of the transverse vibrations of a uniform beam pinned at both ends. Let $w(p,t)$ be the deflection at abscissa p at time t, and assume r_d point force disturbances at abscissae p_i, $i = 1,\dots,r_d$, and r piezoelectric plates bonded on the beam at abscissae p_i, $i = r_d+1,\dots,r_d+r$ to be used as control inputs. Moreover, m velocity sensors are located at abscissae $\bar{p}_i$, $i = 1,\dots,m$, as shown in Figure 4.5

The equation of the beam is (see Equation 2.38)

$$\frac{\partial^2}{\partial p^2}\left[EI(p)\frac{\partial^2 w(p,t)}{\partial p^2}\right] + m(p)\frac{\partial^2 w(p,t)}{\partial t^2} = \sum_{i=1}^{r+r_d} f_i(p,t), \tag{4.30}$$

with boundary conditions

[4] Obviously, extending SISO control strategies to MIMO plants is a frequent practice, as in the case of sequential loop closure [19], but generally the inherent simplicity of the approach is lost, along with the generality of the controller, which reduces to a diagonal matrix of transfer functions, as opposed to a full matrix.

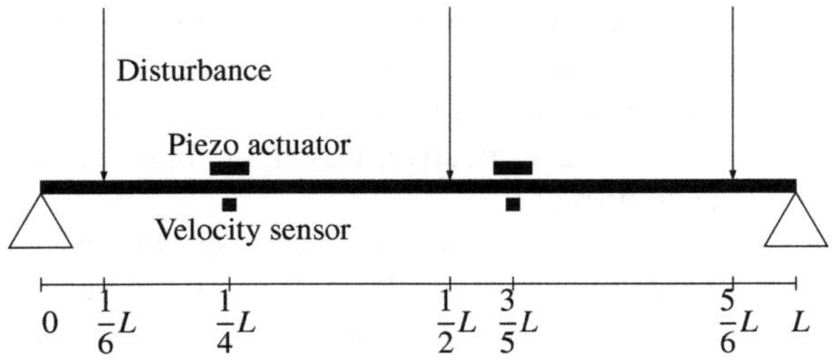

Figure 4.5 Flexible beam configuration

$$w(0,t) = 0, \tag{4.31}$$

$$w(L,t) = 0, \tag{4.32}$$

$$\left.\frac{\partial^2 w(p,t)}{\partial p^2}\right|_{p=0} = 0, \tag{4.33}$$

$$\left.\frac{\partial^2 w(p,t)}{\partial p^2}\right|_{p=L} = 0, \tag{4.34}$$

where L is the beam length, $m(p)$ the mass *per* unit of length, $I(p)$ the moment of inertia and E Young's modulus. The (distributed) forces acting on the beam are denoted by $f_j(p,t)$.

As discussed in Chapter 2, using the modal approach, the solution is decomposed as

$$w(p,t) = \sum_{i=1}^{\infty} W_i(p)\eta_i(t), \tag{4.35}$$

where $W_i(p)$ and $\eta_i(t)$ are the *normalised shape* and the *modal coordinate* of mode i, respectively. Considering also a damping ratio ζ_i for the mode with natural frequency ω_{n_i} and truncating the mathematical description to the first n modes, the dynamics of the structure is described by the following equations:

$$\ddot{\eta}_i(t) + 2\zeta_i\omega_{n_i}\dot{\eta}_i(t) + \omega_{n_i}{}^2\eta_i(t) = \sum_{j=1}^{r+r_d} N_{ij}(t), \quad i = 1,...,n. \tag{4.36}$$

The terms $N_{ij}(t)$ are the projection of $f_j(p,t)$ on the ith mode. As the disturbances are forces concentrated at abscissa p_i, $i = 1,\ldots,r_d$, $N_{ij}(t)$ can be written as

$$N_{ij}(t) = W_i(p_j)d_j(t), \quad j = 1,\ldots,r_d, \tag{4.37}$$

where $d_j(t)$ is the time behaviour of the disturbance [20]. For the piezoelectric actuators, assuming the ends of the jth piezo are at abscissae p_j^- and p_j^+, it is possible to write [7]

$$N_{ij}(t) = \left(\frac{\mathrm{d}W_i(p_j^+)}{\mathrm{d}p} - \frac{\mathrm{d}W_i(p_j^-)}{\mathrm{d}p} \right) u_j(t), \quad j = r_d + 1, \dots, r_d + r. \tag{4.38}$$

The above description is linear and can be cast into a space state form. When considering output feedback, a description of the measured output must also be given. Assuming velocity measurements at abscissae $\bar{p}_i$, the sensor measurements are

$$y_i(t) = \sum_{j=1}^{n} V_j(\bar{p}_i)\dot{\eta}_j(t), \quad i = 1, \dots, m. \tag{4.39}$$

Defining the state vector $x = (\eta_1 \;\; \dots \;\; \eta_n \;\; \dot{\eta}_1 \;\; \dots \;\; \dot{\eta}_n)^{\mathrm{T}}$, the control vector $u = (u_1 \;\; \dots \;\; u_r)^{\mathrm{T}}$, the disturbance vector $d = (d_1 \;\; \dots \;\; d_{r_d})^{\mathrm{T}}$ and the output vector $y = (y_1 \;\; \dots \;\; y_m)^{\mathrm{T}}$, Equations 4.36 and 4.39 can be rewritten as

$$\dot{x} = Ax + B_d d + B_u u, \tag{4.40}$$

$$y = Cx, \tag{4.41}$$

where

$$A = \begin{pmatrix} 0_{n\times n} & I_n \\ -\Omega & -\Lambda \end{pmatrix}, \tag{4.42}$$

$$B_d = \begin{pmatrix} 0_{n\times r_d} \\ B_{d2} \end{pmatrix}, \tag{4.43}$$

$$B_u = \begin{pmatrix} 0_{n\times r} \\ B_{u2} \end{pmatrix}, \tag{4.44}$$

$$C = \begin{pmatrix} 0_{m\times n} & C_2 \end{pmatrix}, \tag{4.45}$$

where Λ and Ω are diagonal matrices.

In order to dampen vibrations of the structure, a *sliding manifold* can be defined such that if the state belongs to the manifold the structure has well damped modes. Thus the control objective becomes to steer the state towards the manifold and to keep it on the manifold (or in a “reasonable” neighbourhood of it) in spite of disturbances.

The sliding manifold can be defined as

$$\mathscr{S} = \left\{ x \in \mathbb{R}^{2n} : s(x) = 0 \right\}, \tag{4.46}$$

where the function $s(x)$ is defined by

$$s(x) = -Hx. \tag{4.47}$$

The selection of the $r \times 2n$ matrix H is based on the following theorem, which is derived from Theorem 1 in [5].

Theorem 4.1. *Consider the system of differential equations*

$$\dot{x} = Ax + B_d d + B_u u, \quad x(0) = 0, \tag{4.48}$$

$$\varepsilon \dot{u} = \dot{s}, \quad u(0) = u_0, \tag{4.49}$$

with the disturbance $d \in \mathscr{L}_\infty$ *with norm* $\|d\|_\infty < \delta$ *and let* γ *and* ν *be positive numbers. Assume moreover that*

$$\operatorname{Re}\lambda_{\min}(HB_u) > \gamma, \tag{4.50}$$

$$\operatorname{Re}\lambda_{\min}(B_u(HB_u)^{-1}HA - A) > \gamma + \nu, \tag{4.51}$$

where $\operatorname{Re}\lambda_{\min}(X)$ *denotes the minimum of the real part of the eigenvalues of the matrix* X.

Then, there exists $\varepsilon_0 > \varepsilon_1 > 0$ *such that for every* $\varepsilon \in (\varepsilon_1, \varepsilon_0]$ *the solution* $(x(t,\varepsilon), u(t,\varepsilon))$ *of the system Equations 4.48 and 4.49 is such that*

$$\|x(t,\varepsilon)\| \leq \kappa\left(1 - e^{-\gamma t}\right)\delta, \tag{4.52}$$

where κ *is a positive constant depending on the data and on the norm of the projection of* B_d *on the complement of the range space of* B_u, i.e., $\|(I - B_u(B_u^{\mathrm{T}}B_u)^{-1}B_u^{\mathrm{T}})B_d\|$ *and the control* u *is*

$$u(t,\varepsilon) = -\frac{1}{\varepsilon}Hx + u_0. \tag{4.53}$$

Proof. Letting $\varepsilon = 0$ in Equation 4.49 the *equivalent control* $\tilde{u}$ is obtained as

$$\tilde{u} = -(HB_u)^{-1}(HA\tilde{x} + HB_d d). \tag{4.54}$$

Thus, according to the singular perturbation theory, condition Equation 4.50 guarantees the stability of the "fast" subsystem. Replacing the equivalent control in Equation 4.48 the *reduced-order system* equation

$$\dot{\tilde{x}} = \left(A - B_u(HB_u)^{-1}HA\right)\tilde{x} + \left(I - B_u(HB_u)^{-1}H\right)B_d d, \quad \tilde{x}(0) = 0, \tag{4.55}$$

is obtained and condition Equation 4.51 proves the stability of the "slow" dynamics. Note that the matrix $\left(I - B_u(HB_u)^{-1}H\right)$ is an annihilator of any vector in the range space of B_u; hence only disturbances in the orthogonal complement of the range space of B_u can affect the reduced-order system. From the bound $\|e^{Xt}\| \leq ke^{(\alpha+\varepsilon)t}$, where $Re\lambda_{\max}X \leq \alpha$ and $\varepsilon > 0$ and $k > 0$ is a constant depending on the norm and the size of the matrix X [6], the bound on the norm of the reduced-order state is deduced. Finally, by resorting to the singular perturbation theory on the infinite horizon [13, 15] the thesis is proved. □

Basically, Theorem 4.1 guarantees stability of the closed loop system. Moreover, disturbances in the range of the input matrix are asymptotically rejected and, by reducing the parameter ε, that is by increasing the control gain, the rejection is more and more effective. If initial conditions different from zero are present, a simple high gain control would result in too high control levels; however, variations of the theorem are possible to take into account initial conditions and limited control rates [5]. Moreover, the theorem says nothing about disturbances in the complement of the input matrix range. However, noting that conditions 4.50 and 4.51 only partially constrain the matrix H, further improvements can be sought. The idea is to use the remaining degrees of freedom to increase the damping of some selected structural modes. This can be done by designing the sliding surface so that r modes are damped out. In order to do this, the state is reordered so that the first $2r$ elements are those related to the modes to be rejected. Letting $\bar{x}$ be the state thus reordered, the dynamic equations are

$$\dot{\bar{x}} = \begin{pmatrix} 0_{r\times r} & I_r & 0_{r\times(n-r)} & 0_{r\times(n-r)} \\ -\Omega_1 & -\Lambda_1 & 0_{r\times(n-r)} & 0_{r\times(n-r)} \\ 0_{(n-r)\times r} & 0_{(n-r)\times r} & 0_{(n-r)\times r} & I_{n-r} \\ 0_{(n-r)\times r} & 0_{(n-r)\times r} & -\Omega_2 & -\Lambda_2 \end{pmatrix} \bar{x} + \begin{pmatrix} 0_{r\times r} \\ \bar{B}_{u1} \\ 0_{(n-r)\times r} \\ \bar{B}_{u2} \end{pmatrix} u. \tag{4.56}$$

Reordering is obtained with the state transformation $\bar{x} = Jx$, where

$$J = \begin{pmatrix} I_r & 0_{r\times(n-r)} & 0_{r\times r} & 0_{r\times(n-r)} \\ 0_{r\times r} & 0_{r\times(n-r)} & I_r & 0_{r\times(n-r)} \\ 0_{(n-r)\times r} & I_{n-r} & 0_{(n-r)\times r} & 0_{(n-r)\times(n-r)} \\ 0_{(n-r)\times r} & 0_{(n-r)\times(n-r)} & 0_{(n-r)\times r} & I_{n-r} \end{pmatrix}. \tag{4.57}$$

Moreover, it is convenient to define the matrix D that transforms the submatrix related to the first r modes to real diagonal form, *i.e.*,

$$D \begin{pmatrix} 0 & I_r \\ -\Omega_1 & -\Lambda_1 \end{pmatrix} D^{-1} = \begin{pmatrix} \alpha & \beta \\ -\beta & \alpha \end{pmatrix}, \tag{4.58}$$

where the first r modes of the matrix A have poles $s_i = -\alpha_i \pm j\beta_i$, $i = 1,\ldots,r$, and $\alpha = \mathrm{diag}(\alpha_i)_{i=1,\ldots,r}$, $\beta = \mathrm{diag}(\beta_i)_{i=1,\ldots,r}$. Finally, it is easy to show that, based on the structure of the matrix A, the matrix D has a special structure

$$D = \begin{pmatrix} \delta_1 & \delta_2 \\ \delta_3 & \delta_4 \end{pmatrix}, \tag{4.59}$$

where $\delta_i = \mathrm{diag}(\delta_{ij})_{j=1,\ldots,r}$, $i = 1,\ldots,4$.

The possibility of choosing the sliding surface so as to reject only r modes while leaving unaffected the remaining $(n-r)$ modes is stated by the following theorem.

Theorem 4.2. *Assume that B_1 is non-singular and select the matrix H in Equation 4.47 as*

$$H = M \begin{pmatrix} \Pi_1 & \Pi_2 \end{pmatrix} \begin{pmatrix} D & 0_{2r\times 2(n-r)} \end{pmatrix} J, \tag{4.60}$$

where M is a square $r \times r$ non-singular matrix, Π_1 and Π_2 are diagonal matrices with elements $\pi_{ij} \in \{-1, 1\}$, $i = 1,2$, $j = 1,\ldots,r$. Then it is possible to choose M, Π_1 and Π_2 to satisfy Theorem 4.1 and increase the damping of the first r modes up to unity.

Proof. Only a sketch of the proof is given here. Details of the proof can be found in [5]. Computing HB_u one has

$$HB_u = M(\Pi_1\delta_2 + \Pi_2\delta_4)B_1, \tag{4.61}$$

since the matrices $(\Pi_1\delta_2 + \Pi_2\delta_4)$ and B_1 are non-singular M can be selected to fulfil condition 4.50. Moreover, by using the change of basis that converts the matrix JAJ to real diagonal form and exploiting the structure of the matrices for flexible systems, it is easy to show that the closed loop dynamic matrix $A - B_u(HB_u)^{-1}HA$ has the last $2(n-r)$ eigenvalues unchanged by the state feedback. Of the remaining $2r$ eigenvalues, r are zero due to the fact that the order of the closed loop system has reduced, as the state trajectory is constrained on the manifold. For the computation of the last r eigenvalues, two cases are possible, *i.e.*, for the ith eigenvalue

1. $|\delta_{2i}| \neq |\delta_{4i}|$, or
2. $|\delta_{2i}| = |\delta_{4i}|$.

In the first case, the closed loop ith eigenvalue is

$$\lambda_i = -\alpha_i - \pi_{1i}\pi_{2i}\beta_i + 2\beta_i \frac{\delta_{2i}\pi_{2i}}{\delta_{2i}\pi_{1i} + \delta_{4i}\pi_{2i}}; \tag{4.62}$$

in the second case, choosing

$$\pi_{1i} = \begin{cases} \pi_{2i} \text{ if } & \delta_{2i} = \delta_{4i} \\ -\pi_{2i} \text{ if } & \delta_{2i} = -\delta_{4i} \end{cases}, \tag{4.63}$$

the ith eigenvalue is

$$\lambda_i = -\alpha_i. \tag{4.64}$$

Thus closed loop stability of the reduced-order system can always be satisfied. □

Note that according to the proof, the matrix M is used only to impose the stability of the fast dynamics. Moreover, although $n-r$ modes are unaffected by the control, the damping of r modes is increased (possibly up to unity). Obviously, the ordering of the modes is arbitrary and hence the strategy is simply to use r inputs to reduce the resonances of r modes, *e.g.*, those with the lowest damping, or those closer to disturbance spectra, if the disturbance is narrow-band.

Obviously, the strategy must be completed with some sort of robust observer, as shown in the following, but first an example is in order to show the effectiveness of the strategy.

4.4.1 Case Study: State Feedback

Consider the beam pinned at both ends in Figure 4.5 with the following data: $L = 3\,\text{m}$, $m(p) = 19.61\,\text{kg/m}$, $I = 6.28 \cdot 10^{-7}\,\text{m}^4$, $E = 2.07 \cdot 10^{11}\,\text{N/m}^2$, $h_p = 30\,\text{mm}$, $\zeta_i = 0.01$, $i = 1,\ldots,4$, with $h_p = p_j^+ - p_j^-$ the length of piezoelectric plates. A simple model with four modes is obtained. Assuming narrow-band disturbances with spectra close to the first two modes, the objective is to increase the damping of

Table 4.1 Open and closed loop modes

Open loop		Reduced-order system		Closed loop system $\varepsilon = 0.001$	
Freq.	Damping	Freq.	Damping	Freq.	Damping
$8.93 \cdot 10$	10^{-2}	$8.93 \cdot 10$	1	$1.03 \cdot 10^2$	$1.53 \cdot 10^{-1}$
$3.57 \cdot 10^2$	10^{-2}	$3.57 \cdot 10^2$	1	$3.89 \cdot 10^2$	$9.31 \cdot 10^{-2}$
$8.04 \cdot 10^2$	10^{-2}	$8.04 \cdot 10^2$	10^{-2}	$8.04 \cdot 10^2$	10^{-2}
$1.43 \cdot 10^3$	10^{-2}	$1.43 \cdot 10^3$	10^{-2}	$1.43 \cdot 10^3$	10^{-2}

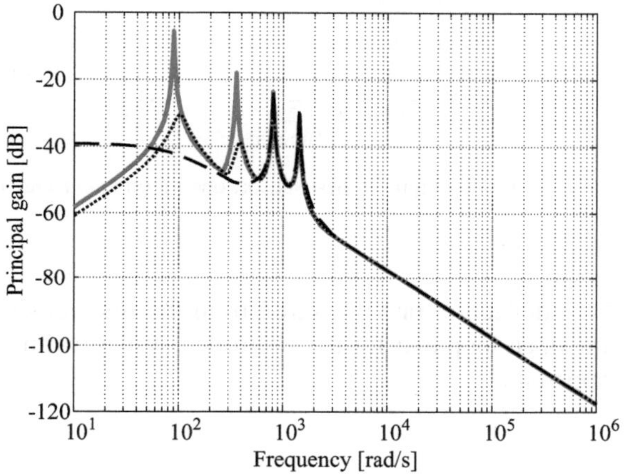

Figure 4.6 Open and closed loop singular values: open loop (*solid grey line*), closed loop (*dotted black line*), reduced order (*dashed black line*)

the first two modes. Choosing

$$M = \begin{pmatrix} -59 & -137 \\ -59 & 137 \end{pmatrix}, \quad \Pi_1 = -I, \quad \Pi_2 = I, \tag{4.65}$$

one has the result in Table 4.1.

In Figure 4.6 the behaviour of the largest singular values of the open loop system is depicted and compared with that of the reduced-order system and the one obtained with the parameter $\varepsilon = 0.001$.

Next, three narrow band disturbances are considered, with spectra close to the first two modes

$$d_1(t) = 30(1 + 0.1 \sin 0.2t) \sin 85t + \sin(320t + \sin 0.1t), \tag{4.66}$$

$$d_2(t) = 30(1 + 0.1 \sin 0.2t) \sin(85t + \pi/2) + \sin(320t + \sin 0.1t), \tag{4.67}$$

$$d_3(t) = 30(1 + 0.1 \sin 0.2t) \sin(85t + \pi/4) + \sin(320t + \sin 0.1t + \pi/3). \tag{4.68}$$

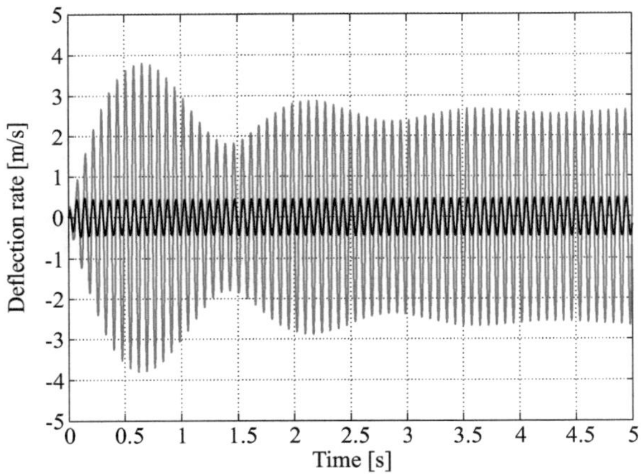

Figure 4.7 Open and closed loop output on first sensor: open loop (*solid grey line*), closed loop (*solid black line*)

In Figure 4.7 the deflection rate measured from the first velocity sensor is reported both for the open loop and for the controlled system, with $\varepsilon = 0.001$. The improvement is apparent.

4.4.2 State Observer

As stated earlier, the control law presented up to this point is a state feedback that cannot be directly implemented on a system with a theoretically infinite number of modes. Moreover, there are no "modal" sensors. Thus a (robust) state observer must be designed; for instance a sliding observer is possible [4, 5].

The equation of the observer is

$$\dot{\hat{x}} = (A - G(q)C)\hat{x} + B_u u + G(q)\left(y - e^{Lt} y_0\right), \tag{4.69}$$

where $\hat{x}$ is the estimated state, L is a Hurwitz matrix and the matrix gain $G(q)$, as in the LTR design [9], depends on a free parameter $q > 0$ and is obtained from

$$G(q) = P(q)C^{\mathrm{T}}R^{-1}, \tag{4.70}$$

and $P(q)$ is the positive definite solution of the the algebraic Riccati equation (ARE)

$$AP(q) + PA^{\mathrm{T}}(q) + Q(q) - P(q)C^{\mathrm{T}}R^{-1}CP(q) = 0, \tag{4.71}$$

with $R > 0$ (*i.e.*, positive definite) and $Q(q) = Q_0 + q^2 B_u V B_u^{\mathrm{T}}$, $Q_0 \geq 0$ and $V > 0$ as in the original formulation of the LTR observer [9].

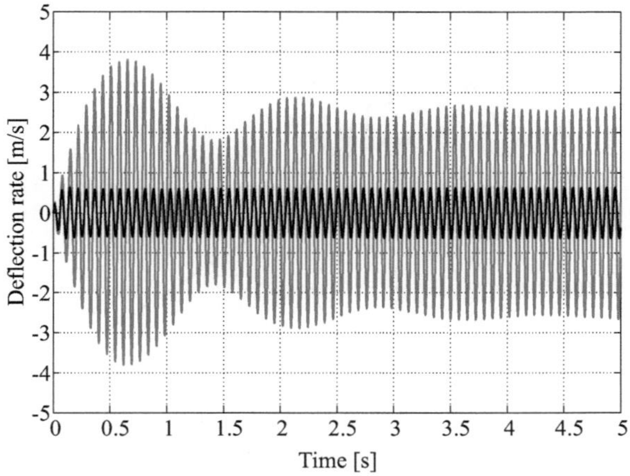

Figure 4.8 Open and closed loop output on first sensor, with state observer: open loop (*solid grey line*), closed loop (*solid black line*)

Using the parameters $q = 10^3$, $R = 0.1I_2$, $V = I_2$, $Q_0 = I_2$, considering $y_0 = 0$ for the sake of simplicity, and with the control law

$$u = -\frac{1}{\varepsilon}H\hat{x}, \tag{4.72}$$

the results in Figure 4.8 are obtained, with the three disturbances in Equations 4.66–4.68.

It is apparent that the use of the observer has introduced only a slight degradation of performances. However, what can really be harmful for the design is the possible presence of spillover. It is indeed clear that, although the state feedback can not introduce spillover, as it only works on modelled dynamics, the use of the observer introduces the risk of spillover, and this explains why it is of paramount importance to use *robust* observers. The idea is to consider unmodelled dynamics as uncertainties, thus the use of robust techniques means filtering out unmodelled dynamics.

In order to check the robustness of the presented strategy, a high-order model has been deduced (8th order), but the same controller as before has been used to test its robustness. The results are reported in Figure 4.9, showing the good performances in the frequency domain and in Figure 4.10 for the time domain.

The results show good performances for the observer-based implementation of the control strategy. It has again to be remarked that this strategy uses r inputs to increase the damping of r modes. This explains why it can be used only for low-modal density structures: high-modal densities would require too many inputs.

For this reason a completely different strategy will be proposed in the next chapter to deal with high-modal density structures.

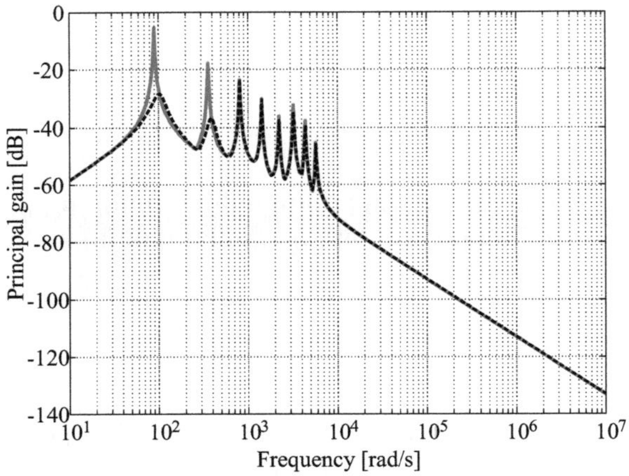

Figure 4.9 Open and closed loop singular values, with robust observer in the perturbed case: open loop (*solid grey line*), closed loop (*dotted black line*)

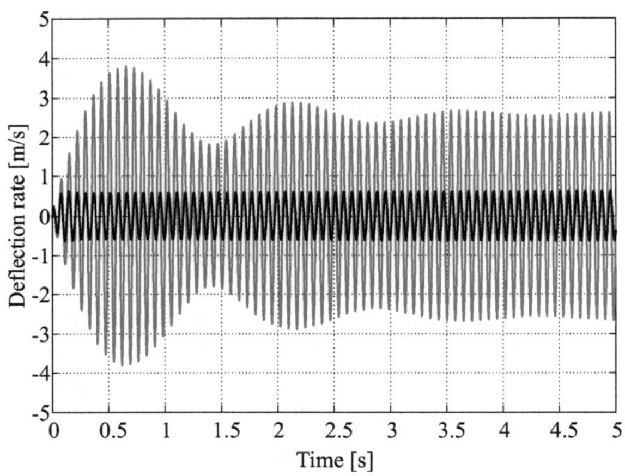

Figure 4.10 Open and closed loop time history for the first channel, with robust observer in the perturbed case: open loop (*solid grey line*), closed loop (*solid black line*)

4.5 Conclusions

Initially, the chapter has presented a classification of the control problem for flexible systems based both on the input (disturbance) signal and on the structure to be controlled. Next, the spillover phenomenon has been presented, which is a typical problem to face in active control of flexible structures. Then, two different approaches, *i.e.*, the internal model-based control and the modal control, have been presented as solutions to active vibration reduction for structures with low modal density and/or subject to tonal (narrow-band) disturbances. The techniques presented in this chapter, together with the optimal control strategies that will described in Chapter 5, provide the reader with a comprehensive overview of all available techniques for the active feedback vibration control. Finally, the chapter ended with some simulations of the described control techniques applied to a case study, *i.e.*, a uniform beam pinned at both ends.

References

1. Balas MJ (1982) Trends in large space structure control theory: fondest hopes, wildest dreams. IEEE Trans Autom Contr, 27:522–535
2. Balas MJ (1978) Feedback Control of Flexible Systems. IEEE Trans Autom Contr, 23:673–679
3. Burke SE, Hubbard JE Jr (1987) Active Vibration Control of a Simply Supported Beam Using a Spatially Distributed Actuator. IEEE Control System Magazine, 8:25–30
4. Cavallo A, De Maria G, Nistri P (1994) Sliding Mode Techniques with Robust Observer. In: Proc. of VSLT94, Benevento (I), 261–265.
5. Cavallo A, De Maria G, Setola R (1999) A Sliding Manifold Approach for the Vibration Reduction of Flexible Systems. Automatica, 35:1689–1696
6. Coppel WA (1978) Dichotomies in Stability Theory. Springer, Berlin
7. Crawley EF, de Luis J (1987) Use of Piezoelectric Actuators as Elements of Intelligent Structures. AIAA Journal, 25:1373–1385
8. Dorf RC, Bishop RH (2001) Modern Control Systems. 9th Ed., Prentice Hall, Upper Saddle River, NJ
9. Doyle J, Stein G (1979) Robustness with Observers. IEEE Trans Autom Contr, 24:607-611
10. Fuller CR, von Flotow AH (1995) Active Control of Sound and Vibrations, IEEE Control Systems Magazine. 15:9–19
11. Fuller CR, Elliot SJ, Nelson PA (1995) Active Control of Vibration. Academic Press, London.
12. Gerhold H (1997) Active Control of Fan-Generated Tone Noise. AIAA Journal, 35:17–22
13. Hoppensteadt FC (1966) Singular Perturbations on the Infinite Interval. Trans Amer Math Soc, 123:521–535
14. Junkins JL, Kim Y (1993) Introduction to Dynamics and Control of Flexible Structures. AIAA Education Series, Washington DC
15. Kokotović PV, Khalil KH, O'Reilly J (1986) Singular Perturbations Methods in Control: Analysis and Design. Academic Press, New York
16. Landau ID, Lozano R, M'Saad M (1998) Adaptive Control. Springer, London
17. Landau ID, Constantinescu A, Rey D (2005) Adaptive Narrow Band Disturbance Rejection Applied to an Active Suspension–An Internal Model Principle Approach. Automatica, 41:563–574
18. Lueg P (1936) Process of Silencing Sound Oscillations. US Patent no. 2,043,416
19. Maciejowski J.M. (1989) Multivariable Feedback Design. Addison-Wesley, Wokingham
20. Meirovitch L. (1990) Dynamic and Control of Structures. Wiley-Interscience, New York
21. Meurers T, Veres SM, Tan ACH (2003) Model-free Frequency Domain Iterative Active Sound and Vibration Control. Contr Eng Pract, 11:1049–1059
22. Madkour A, Hossain MA, Dahal KP, Yu H (2007) Intelligent Learning Algorithms for Active Vibration Control. IEEE Trans Syst Man Cybern C Appl Rev, 37:1022–1033
23. Olgac N, Hosek M (1997) Active Vibration Absorption Using Delayed Resonator with Relative Position Measurement. ASME J Vib Acoust, 119:131-136
24. Widrow B, Stearns SD (1985) Adaptive Signal Processing. Prentice-Hall, Englewood Cliffs, NJ
25. Wie B, Gonzales M (1992) Control Synthesis for Flexible Space Structures Excited by Persistent Disturbances. J Guidance, 15:73–80
26. Youla DC, Bongiorno JJ, Lu CN (1974) Single-loop Stabilization of Linear Multivariable Plants. Automatica, 10:159–173

Chapter 5
Optimal Feedback Control of Flexible Structures

5.1 Introduction

In the previous chapter the problem of controlling low modal density structures or systems affected by narrow-band disturbances was addressed. At the other end of the spectrum is the problem of broadband disturbance rejection for systems with harmonic response not exhibiting the classical "spikes" of systems with low modal density. Typical applications are aeronautic and aerospace ones, where the need for light materials and devices is leading towards the adoption of structures where elastic phenomena are emphasised, with complex geometries used to increase the mechanical stiffness of the structures. At the beginning of the 1990s the availability of new piezoelectric devices, and the increased interest in challenging space programs (*e.g.*, the Spacelab) focussed the attention of the researchers on the control of plate structures [11].

For this kind of application the basic approach is to use colocated velocity feedback, with piezoelectric devices used as actuators and velocity measurements obtained by filtering (integrating) the signal resulting from accelerometers, so as to guarantee minimum phase and real-positiveness properties, at least in a suitable frequency range, *i.e.*, up to frequencies with wavelength comparable to the size of piezo actuators. In theory these properties allow the use of high-gain control strategies, with the good characteristics in terms of disturbance rejection and robustness inherent in this kind of control. However, the presence of unmodelled dynamics prevents the designer from increasing the gains too much.

It has been considered that a simple velocity feedback is unable to selectively reduce vibrations [11], since it does not distinguish among different frequencies. Simple frequency-selective capabilities can be inserted by using filters in order to change the loop gain over different frequencies, but this also induces phase lags, that are harmful especially near the cross-over frequency. Thus the only practical solution with this simple approach is to keep the gains low, accepting the drawback of degraded performances.

A more sophisticated solution would be to use optimal control techniques, like $\mathscr{H}_\infty$ or μ-synthesis. This would also have beneficial effects on robustness. Indeed, it has been shown [2] that poor uncertainty modelling can degrade closed loop performances when implemented experimentally, leading even to instability. The problem is that often the model used in control design is unreliable not only in high frequencies, but also at *low frequency*, where some sensors (*e.g.*, accelerometers) used in the preliminary identification phase have a low signal-to-noise ratio.

Fortunately, this selective reliability of the model is not in conflict with the problem specifications. In fact in the active control of sound and vibration fields high frequency (greater than 1 kHz) vibrations are usually reduced *via* passive techniques [12], while active control is effective for *relatively* low frequency. The term *relatively* deserves some comments: in practical applications control efforts at *very low frequency* are often useless or cannot be implemented in feedback, as is the case with rigid mode dynamics, which obviously cannot be controlled by piezoelectric devices. In this case, also low frequencies must be filtered out, otherwise the very low frequency components of the measured signal would saturate the actuators, thus resulting in a very poor control performance. Moreover, accelerometers suffer from offsets due to mechanical and thermal effects [9]. Thus vibration reduction specifications are generally assigned only in a prescribed frequency band, *e.g.*, in the range 50 to 1000 Hz [12].

When modal density is moderate some *ad hoc* solutions can be employed, like placing actuators at some nodal points of the structure, thus making uncontrollable selected modes and increasing the frequency distance between controlled modes and residual ones [16]. However, if the modal density is high a different approach must be pursued, as will be illustrated next. Preliminarily, other constraints of the problem have to be highlighted.

One more constraint to the design is the need for a stable stabilising controller. In fact, using unstable controllers makes implementation extremely difficult, since only when the control loop is closed and effective is there a guarantee that all the signals in the loop are limited. However, during the loop closure, the output of the controller increases and this may saturate the actuators. This implementation argument against unstable controllers is reinforced by one more reason, a theoretical one: it is known that unstable poles in the forward path of the control loop impose a minimum bandwidth [17, 23]. This contrasts with the bandwidth limitation imposed by rejection of unmodelled high frequency modes, in order to avoid spillover and poor robustness, as mentioned above. Note that automated design procedures, such as optimal control, generally only produce *stabilising* controllers, not necessarily stable ones. Only *strongly stabilising* techniques produce stable controllers. Stable stabilising controllers have been discussed in recent years, starting from the seminal paper by Youla *et al.* [20], which showed that the parity interlacing property (p.i.p.) is necessary and sufficient for the strong stabilisation. Based on this result, several papers have been written addressing the problem of checking the p.i.p. without computing the zeros of the system (*e.g.*, see [3] for a Routh table-based test). Once the p.i.p. is satisfied, only the existence of a stable controller is guaranteed, and the task of the control designer becomes to find it. The problem is all but trivial, especially in the multivariable case, since in many situations standard techniques (*e.g.*, $\mathscr{H}_2$, $\mathscr{H}_\infty$) can lead to unstable controllers, depending on the selection of the performance weights. A very effective solution appeared in 1994 [14], when Halevi proposed a suitable selection of the weighting matrices for a linear quadratic Gaussian (LQG) problem so as to achieve strong stability. In [1] a review of different lines of attack to the problem are presented, from nonlinear programming to Halevi-like solutions. Since mid 1990s, also strong stabilising $\mathscr{H}_\infty$ controllers have been sought for by using interpolation conditions [21]. Also, by using the free parameter of the parametrisation of all $\mathscr{H}_\infty$ controllers guaranteeing a prescribed closed loop $\mathscr{H}_\infty$ norm, a stable controller [22, 23] with maximum order equal to twice the order of the plant has been found. Sufficient conditions requiring the solution of two coupled Riccati equations and a spectral radius inequality have also been provided [5], as well as solutions based on linear matrix inequalities (LMIs) [13] and on parametric optimisation [4].

An interesting property exhibited by flexible systems with colocated actuator-sensor pairs is that they are known to possess the p.i.p., even in the multivariable case [6], thus this line of attack seems encouraging. A very recent solution was proposed in [7], where a suitable closed-form of an $\mathscr{H}_\infty$ was proposed, addressing also the above constraints. This solution will be presented and discussed in the rest of the chapter.

In particular, the presented controller is (1) stabilising, (2) stable, (3) frequency-selective, (4) has transmission zeros at the origin so as to have bandpass capabilities and (5) optimises an $\mathscr{H}_\infty$ criterion, thus reducing the peaks in the closed loop system in a continuous frequency range. Specifically, the optimised index is based on a combination of vibration reduction at the most sensitive frequencies of the system, *i.e.*, those corresponding to natural modes of vibration of the structure, and robustness against matched disturbances (*i.e.*, those in the range of the control input matrix). Furthermore, by resorting to the properties of the flexible structure, the proposed technique results in a closed-form solution to an $\mathscr{H}_\infty$ control problem, thus avoiding the need to solve any Riccati equations. This makes the strategy very attractive for high-dimensional or ill-conditioned control problems, where numerical solvers of the Riccati equation may fail. Moreover, exploiting the particular structure of the considered plant and of the proposed solution, also an interpretation in the framework of the internal model principle is given of the $\mathscr{H}_2$ controller, and the explicit solution for the parametrisation of all stabilising $\mathscr{H}_\infty$ and $\mathscr{H}_2$ controllers is given and a whole *family* of controllers with the above properties is presented. The effectiveness and the robustness of the proposed solution are shown *via* a numerical simulation on an eighth-order flexible system, where up to 50% uncertainties in the nominal parameters are considered. Experimental testing is left to Chapter 7.

5.2 Control Model

Again, it is assumed that a modal approach has been used to deduce the model of the structure, according to the methodology presented in Chapters 2 and 3. The dynamics of the flexible structure is thus approximated by considering n modes, resulting in the set of equations

$$\ddot{\eta}_i(t) + 2\zeta_i \omega_{n_i} \dot{\eta}_i(t) + {\omega_{n_i}}^2 \eta_i(t) = F_i(t),\ i = 1,\ldots,n, \tag{5.1}$$

where $\eta_i(t), i = 1,\ldots,n$ are the *modal coordinates* with natural frequency and damping coefficient ω_{n_i} and ζ_i, respectively, and $F_i(t)$ are the *modal forces*.

In this section m point actuators and m point sensors are considered. "Square" systems (*i.e.*, with as many inputs as outputs) are used at this point only for the sake of simplicity in the presentation, and the square assumption will be removed later. The problem is cast into a TITO (two-input-two-output) framework, shown in Figure 5.1, where $\tilde{P}$ is the system to be controlled

$$\begin{pmatrix} \tilde{z} \\ \tilde{y} \end{pmatrix} = \begin{pmatrix} \tilde{P}_{11} & \tilde{P}_{12} \\ \tilde{P}_{21} & \tilde{P}_{22} \end{pmatrix} \begin{pmatrix} \tilde{w} \\ \tilde{u} \end{pmatrix} = \tilde{P} \begin{pmatrix} \tilde{w} \\ \tilde{u} \end{pmatrix} \tag{5.2}$$

and $\tilde{K}$ is the controller, with the system exhibiting two inputs, an unmanipulable one, and a control input, and two outputs, a performance output and a control output. The objective is to find a controller to be connected between control input and measured outputs in order to minimise the effect of unmanipulable inputs on performance outputs. Moreover, state disturbances and measurement noise have to be considered for a realistic design. Let $x(t) = (\,\eta_1(t) \;\; \ldots \;\; \eta_n(t) \;\; \dot{\eta}_1(t) \;\; \ldots \;\; \dot{\eta}_n(t)\,)^{\mathrm{T}} = (\,x_1^{\mathrm{T}} \;\; x_2^{\mathrm{T}}\,)^{\mathrm{T}}$ denote the state, $\tilde{u} \in \mathbb{R}^m$ the control input, $\tilde{y} \in \mathbb{R}^m$ the measured output, $\tilde{w} \in \mathbb{R}^{n+m}$ the generalised disturbance, *i.e.*, state disturbances and measurement noise, $\tilde{z} \in \mathbb{R}^{n+m}$ the performance output, then the model $\tilde{P}$ of the structure is described by

$$\dot{x} = Ax + \tilde{B}_w \tilde{w} + \tilde{B}_u \tilde{u}, \tag{5.3}$$
$$\tilde{z} = \tilde{C}_z x \qquad\qquad + \tilde{D}_{zu} \tilde{u}, \tag{5.4}$$
$$\tilde{y} = \tilde{C}_y x + \tilde{D}_{yw} \tilde{w}, \tag{5.5}$$

where

$$A = \begin{pmatrix} 0 & I \\ -\Omega & -\Lambda \end{pmatrix}, \quad \begin{aligned} \Omega &= \mathrm{diag}(\omega_{n_1}^2, \ldots, \omega_{n_n}^2), \\ \Lambda &= \mathrm{diag}(2\zeta_1 \omega_{n_1}, \ldots, 2\zeta_n \omega_{n_n}), \end{aligned} \tag{5.6}$$

$$\tilde{B}_u = \begin{pmatrix} 0 \\ \tilde{B}_{u2} \end{pmatrix}, \quad \tilde{C}_y = \begin{pmatrix} 0 & \tilde{B}_{u2}^{\mathrm{T}} \end{pmatrix}, \tag{5.7}$$

and $\tilde{B}_{u2} \in \mathbb{R}^{n \times m}$ is a full-column rank matrix. The transfer function from the generalised disturbance $\tilde{w}$ to the performance output $\tilde{z}$ is given by the lower linear fractional transformation (LFT) $\tilde{F}_l(\tilde{P}, \tilde{K}) = \tilde{P}_{11} + \tilde{P}_{12}\tilde{K}\left(I - \tilde{P}_{22}\tilde{K}\right)^{-1}\tilde{P}_{21}$, and denoted by $T_{\tilde{z}\tilde{w}}$.

Moreover, $\tilde{D}_{zu}$ and $\tilde{D}_{yw}$ are assumed full-column rank and full-row rank, respectively. It is well known [17] that the above hypothesis implies that scaling transformations can be found such that the $\tilde{D}$ matrices assume a simpler (normalised) structure. In fact, consider nonsingular transformations S_u and S_y, and unitary trans-

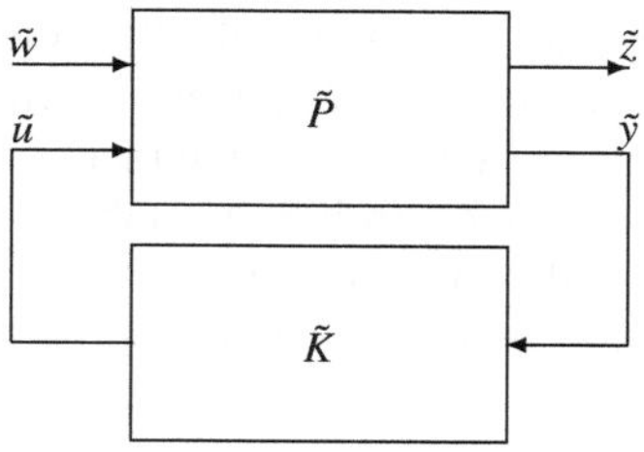

Figure 5.1 Standard control problem

formations Q_w, Q_z, all of suitable dimensions, and define scaled variables w, u, z and y by

$$\tilde{u} = R_u u, \tilde{y} = R_y y, \tilde{w} = Q_w w, \tilde{z} = Q_z z. \tag{5.8}$$

Then, it is possible to choose the scaling matrices R_y, R_u, Q_z, Q_y such that

$$\begin{aligned} D_{zu} &= Q_z^{\mathrm{T}} \tilde{D}_{zu} R_u = \begin{pmatrix} 0 \\ I \end{pmatrix}, \\ D_{yw} &= R_y^{-1} \tilde{D}_{yw} Q_w = \begin{pmatrix} 0 & I \end{pmatrix}, \end{aligned} \tag{5.9}$$

and the scaled model becomes

$$\dot{x} = Ax + B_w w + B_u u, \tag{5.10}$$
$$z = C_z x \qquad + D_{zu} u, \tag{5.11}$$
$$y = C_y x + D_{yw} w, \tag{5.12}$$

where $B_w = \tilde{B}_w Q_w$, $B_u = \tilde{B}_u R_u = \begin{pmatrix} 0 & B_{u2}^{\mathrm{T}} \end{pmatrix}^{\mathrm{T}}$, $C_z = Q_z^{\mathrm{T}} \tilde{C}_z$ and $C_y = R_y^{-1} \tilde{C}_y$.

Finally, let

$$B_w = \begin{pmatrix} 0 & 0 \\ B_{w2} & 0 \end{pmatrix}, \quad C_z = \begin{pmatrix} 0 & B_{w2}^{\mathrm{T}} \\ 0 & 0 \end{pmatrix}. \tag{5.13}$$

The choice of the disturbance matrix B_w is not inflexible and is usually done on heuristic considerations based on the designer's experience and intuition. However, this fact leaves considerable flexibility in the selection, hence it is possible to think of a different strategy, where the focus is on being able to fulfil the constraints posed on the controller (stability, limited bandwidth), while retaining some physical meaning in the mathematical description of the unmanipulable variables. Another element to be taken into account is the robustness capabilities gained from the approach. In fact, it is well known that in $\mathscr{H}_\infty$-based control, the closed loop system is able to withstand unstructured uncertainties with magnitude bounded by the reciprocal of the T_{zw} infinity norm. The choice of B_w in Equation 5.13 has a physical justification, *i.e.*, the relationship between modal displacement and velocity coordinates is not subject to uncertainties or perturbations, thus the first n rows of B_w are zero. Also the selection of the matrix C_z has a physical interpretation, since in noise and vibration reduction, the focus is on the velocity reduction, rather than on modal coordinates control. Finally, the matrix B_{w2} is to be selected. However, no general results are available on how the selection of the matrix B_{w2} influences the location of the zeros and the poles of the controller. As stability and zeros at the origin are a requirement in the proposed design, a choice of the matrix B_{w2} able to produce a strongly stabilising bandpass controller is now proposed.

5.3 Choice of the Disturbance Matrix

In the following, a suitable disturbance matrix is proposed in order to design a strongly stabilising $\mathscr{H}_\infty$ controller with a set of m zeros at the origin, so as to guarantee bandpass properties to the controller. Let

$$B_{w2} = (\alpha\Lambda + B_{u2}B_{u2}^{\mathrm{T}})^{1/2}, \tag{5.14}$$

where α is a scalar coefficient to be selected. Note that for the real square root matrix to exist, the matrix $\alpha\Lambda + B_{u2}B_{u2}^{\mathrm{T}}$ must be positive semi-definite, which in turn requires $\alpha \geq 0$, as shown by considering the SVD of the matrix $B_{u2}B_{u2}^{\mathrm{T}}$. In practical applications the matrix B_{w2} is often required to be positive definite, which requires $\alpha > 0$.

From the physical point of view, the matrix B_w resulting from Equations 5.13 and 5.14 is composed of two terms: the first weights disturbances in the range of B_u, while the second considers also off-range terms, in particular by weighting the system natural modes. Note that all the modes are weighted, and thus $\mathscr{H}_\infty$ design will try to reduce structural peaks due to vibration modes and to increase robustness with respect to *matched* disturbances (*i.e.*, disturbances in the range of the control input matrix), which is a classic requirement in control system design. Moreover, the choice of the matrix B_w also affects robustness with respect to bounded uncertainties affecting the relationship between the variables z and w, *i.e.*, uncertainties whose "size" is inversely bounded by the infinity norm of the transfer function T_{zw}.

5.4 Bandpass Controllers

Before discussing the solution to the $\mathscr{H}_\infty$ problem, a characterisation of controllers with bandpass capabilities is in order. The following theorem addresses this issue.

Theorem 5.1. *Assume that a realisation of the controller $K(s)$ for the system Equations 5.10–5.12 is*

$$\dot{x}_c = A_c x_c + B_c y, \tag{5.15}$$

$$u = C_c x_c, \tag{5.16}$$

where

$$A_c = A + a\hat{B}\Upsilon^{-1} + b\Upsilon\hat{B}, \tag{5.17}$$
$$B_c = c\Upsilon B_u, \tag{5.18}$$
$$C_c = dB_u^{\mathrm{T}}\Upsilon^{-1}, \tag{5.19}$$

and a, b, $c \neq 0$*,* $d \neq 0$ *are real scalars,* $\hat{B} = B_u B_u^{\mathrm{T}}$ *and* Υ *is a symmetric positive definite matrix. If* Υ *is chosen as*

$$\Upsilon = \begin{pmatrix} \Upsilon_1 & 0 \\ 0 & \Upsilon_2 \end{pmatrix}, \tag{5.20}$$

with $\Upsilon_1, \Upsilon_2 \in \mathbb{R}^{n\times n}$*, then K has m zeros at* $s = 0$*.*

Proof. Since B_u has full-column rank, there exists an orthogonal matrix $Q_2 \in \mathbb{R}^{n\times n}$ such that

$$B_{u2}^{\mathrm{T}} Q_2 = \begin{pmatrix} 0 & S_1 \end{pmatrix}, \tag{5.21}$$

where $S_1 \in \mathbb{R}^{m\times m}$ is nonsingular.

Let $Z_c \in \mathbb{R}^{(2n+m)\times(2n+m)}[s]$ (the ring of square $2n+m$ matrix of polynomials in the variable s) be the Rosenbrock system matrix

$$Z_c = \begin{pmatrix} sI - A_c & B_c \\ C_c & 0 \end{pmatrix}. \tag{5.22}$$

Moreover define unimodular matrices $U \in \mathbb{R}^{2n\times 2n}[s]$

$$U = \Upsilon \begin{pmatrix} I_n & 0 \\ 0 & Q_2 \end{pmatrix} \tag{5.23}$$

and $V \in \mathbb{R}^{(2n+m)\times(2n+m)}[s]$

$$V = \begin{pmatrix} U & 0 \\ 0 & I_m \end{pmatrix}. \tag{5.24}$$

Then $V^{-1}Z_cV$ has the same transmission zeros [10] as Z_c, *i.e.*,

$$V^{-1}Z_cV = \begin{pmatrix} sI - U^{-1}A_cU & \begin{pmatrix} 0_{(2n-m)\times m} \\ cS_1^{\mathrm{T}} \end{pmatrix} \\ \begin{pmatrix} 0_{(2n-m)\times m} & dS_1 \end{pmatrix} & 0 \end{pmatrix}. \tag{5.25}$$

Since S_1 is a square invertible matrix and $c, d \neq 0$ it is possible to eliminate the $(2n-m+1)$th to $2n$th rows and columns with elementary operations, thus reducing the above problem to finding the zeros at the origin of the matrix

$$L_m^{\mathrm{T}} U^{-1}(sI - A_c)UL_m, \tag{5.26}$$

with

$$L_m = \begin{pmatrix} I_{2n-m} \\ 0_{m\times(2n-m)} \end{pmatrix}, \tag{5.27}$$

i.e., the eigenvalues of the matrix $L_m^{\mathrm{T}} U^{-1} A_c U L_m$, that are the eigenvalues of the matrix

$$A_r = L_m^{\mathrm{T}} U^{-1} A U L_m, \tag{5.28}$$

that in turn, due to (5.20), are the eigenvalues of the matrix A_r obtained by extracting the first $2n-m$ rows and columns of the matrix

$$\begin{pmatrix} 0_{n\times n} & \Upsilon_1^{-1}\Upsilon_2 Q_2 \\ -Q_2^{\mathrm{T}}\Upsilon_2^{-1}\Omega\Upsilon_1 & -Q_2^{\mathrm{T}}\Upsilon_2^{-1}\Lambda\Upsilon_2 Q_2 \end{pmatrix}. \tag{5.29}$$

Since the first n columns of the above matrix have full rank due to the non-zero $(2,1)$ block, by removing the last m rows, the rank of the first n columns drops to $n-m$ (at the most), so that the rank of the $(2n-m)\times(2n-m)$ matrix A_r cannot exceed $2n-2m$, thus resulting in m rank deficiency. □

Theorem 5.1 gives only a sufficient condition for the choice of a set of parameters resulting in a controllers with m zeros at the origin. In fact, any choice of the matrix Υ such that the matrix A_r has m eigenvalues in zero would result in the same property. In general, however, this condition is not easy to check. Indeed, based on (5.28), the problem is to find a change of basis on the matrix A such that the first $2n-m$ columns (or rows) of the transformed matrix are not linearly independent and include exactly m linearly dependent columns. However, as a counterexample, it is easy to prove that any matrix Υ of the form

$$\Upsilon = \begin{pmatrix} \Upsilon_1 & \Upsilon_{12} \\ \Upsilon_{12}^{\mathrm{T}} & \alpha I \end{pmatrix},$$

where α is a real scalar, satisfies the same property if $\mathrm{Ker}(\Upsilon_{12}) \supseteq \mathrm{Im}(B_{u2})$. Note that this condition makes the counterexample rather weak, since in Equations 5.17–5.19 the effect is the same as $\Upsilon_{12} = 0$.

5.5 $\mathscr{H}_\infty$ Controller Design

There is another motivation for choosing a block diagonal structure of the Υ matrix, namely the possibility to satisfy suitable AREs that will be needed by the $\mathscr{H}_\infty$ design, as shown by the following lemma.

Lemma 5.1. *Consider the system of Equations 5.10–5.12 and let X_∞ and Y_∞ be $(2n\times 2n)$ positive semi-definite symmetric matrices satisfying*

$$A^{\mathrm{T}} X_\infty + X_\infty A + X_\infty(\gamma^{-2} B_w B_w^{\mathrm{T}} - B_u B_u^{\mathrm{T}}) X_\infty + C_z^{\mathrm{T}} C_z = 0, \tag{5.30}$$

$$A Y_\infty + Y_\infty A^{\mathrm{T}} + Y_\infty(\gamma^{-2} C_z^{\mathrm{T}} C_z - C_y^{\mathrm{T}} C_y) Y_\infty + B_w B_w^{\mathrm{T}} = 0, \tag{5.31}$$

for a given $\gamma > 0$. Then, a necessary condition for the solution to the AREs 5.30 and 5.31 to be of the form

$$X_\infty = \beta^2 Y_\infty^{-1}, \tag{5.32}$$

where $\beta < \gamma$ is a real scalar, is that X_∞ and Y_∞ are block diagonal.

Proof. Assuming that a solution of the form of Equation 5.32 exists, pre- and post-multiplying Equation 5.30 by βX_∞^{-1} and taking into account Equation 5.32, one has

$$Y_\infty A^{\mathrm{T}} + AY_\infty + \beta^2\left(\gamma^{-2} B_w B_w^{\mathrm{T}} - B_u B_u^{\mathrm{T}}\right) + \beta^{-2} Y_\infty C_z^{\mathrm{T}} C_z Y_\infty = 0. \tag{5.33}$$

Comparing Equations 5.33 and 5.31, the following equality results:

$$Y_\infty\left((\beta^{-2} - \gamma^{-2}) B_w B_w^{\mathrm{T}} + B_u B_u^{\mathrm{T}}\right) Y_\infty = \left(1 - (\beta/\gamma)^2\right) B_w B_w^{\mathrm{T}} + \beta^2 B_u B_u^{\mathrm{T}}, \tag{5.34}$$

i.e.,

$$\begin{pmatrix} Y_2 \hat{B} Y_2^{\mathrm{T}} & Y_2 \hat{B} Y_3 \\ Y_3 \hat{B} Y_2^{\mathrm{T}} & Y_3 \hat{Y}_3^{\mathrm{T}} \end{pmatrix} = \begin{pmatrix} 0 & 0 \\ 0 & \beta^2 \tilde{B} \end{pmatrix}, \tag{5.35}$$

where

$$Y_\infty = \begin{pmatrix} Y_1 & Y_2 \\ Y_2^{\mathrm{T}} & Y_3 \end{pmatrix} \tag{5.36}$$

and

$$\hat{B} = (\beta^{-2} - \gamma^{-2}) B_{w2} B_{w2}^{\mathrm{T}} + B_{u2} B_{u2}^{\mathrm{T}}. \tag{5.37}$$

Since $\hat{B} > 0$, the $(1,1)$ entry of the matrix on the left-hand side of Equation 5.35 implies $Y_2 = 0$, *i.e.*, Y_∞ is block diagonal and, due to 5.32, so must be X_∞. □

Based on the above results, an $\mathscr{H}_\infty$ controller is derived with bandpass frequency shape. Moreover, the controller is also stable, as shown in the following theorem, which is the main theorem of this section.

Theorem 5.2. *Consider the system 5.10–5.12, assume the pair (A, B_u) to be stabilisable and the pair (A, C_y) be detectable, and let B_w be chosen as in Equations 5.13 and 5.14. Then, the following controller is strongly stabilising and has m zeros at the origin:*

$$K_\infty(s,\gamma) = -\frac{\gamma^2}{\gamma^2 - 2} B_u^{\mathrm{T}} (sI - A_\infty)^{-1} B_u, \tag{5.38}$$

where

$$A_\infty = \begin{pmatrix} 0 & I \\ -\Omega & -\Lambda_\infty \end{pmatrix} \tag{5.39}$$

and

$$\Lambda_\infty = \left(1 - \frac{2}{\gamma^2}\right)\Lambda + 2\frac{\gamma^2 - 1}{\gamma^2 - 2}\sqrt{\frac{\gamma^2 - 1}{\gamma^2}} B_{u2} B_{u2}^{\mathrm{T}}. \tag{5.40}$$

Moreover, if

$$\alpha = 2\sqrt{1 - \gamma^{-2}} \tag{5.41}$$

is considered in Equation 5.14, then the LFT T_{zw} satisfies

$$\|T_{zw}\|_\infty < \gamma. \tag{5.42}$$

Finally, $\inf \gamma = \sqrt{2}$ *and the controller*

$$K_{\text{opt}} = B_{u2}^{\mathrm{T}}(B_{u2}B_{u2}^{\mathrm{T}})^{\dagger}B_{u2}, \tag{5.43}$$

where $(\cdot)^{\dagger}$ *denotes the pseudoinverse, achieves the optimal value*

$$\min_K \|T_{zw}\|_\infty = \sqrt{2}, \tag{5.44}$$

where K *belongs to the set of stable, stabilising controllers with* m *zeros at* $s = 0$.

Proof. Consider the suboptimal case first. Controllability of the pair (A, B_w) and observability of the pair (A, C_z) result from the positive definiteness of matrix 5.14. Then, it is well known [24] that for the controller K to be a stabilising controller such that the LFT T_{zw} has infinity norm smaller than γ the following conditions must be fulfilled:

(i) There exists a stabilising solution $X_\infty > 0$ such that

$$X_\infty A + A^{\mathrm{T}} X_\infty + X_\infty(\gamma^{-2} B_w B_w^{\mathrm{T}} - B_u B_u^{\mathrm{T}}) X_\infty + C_z^{\mathrm{T}} C_z = 0. \tag{5.45}$$

(ii) There exists a stabilising solution $Y_\infty > 0$ such that

$$A Y_\infty + Y_\infty A^{\mathrm{T}} + Y_\infty(\gamma^{-2} C_z^{\mathrm{T}} C_z - C_y^{\mathrm{T}} C_y) Y_\infty + B_w B_w^{\mathrm{T}} = 0. \tag{5.46}$$

(iii) Let $\rho(X)$ denote the spectral radius of the matrix X, then

$$\rho(X_\infty Y_\infty) < \gamma^2.$$

Moreover, a solution to the AREs in the block-diagonal form of Equation 5.20 is sought, in order to guarantee the "derivative" property of the solution, *i.e.*, solutions of the form

$$X_\infty = \begin{pmatrix} X_{\infty,1} & 0 \\ 0 & X_{\infty,2} \end{pmatrix}, \tag{5.47}$$

$$Y_\infty = \begin{pmatrix} Y_{\infty,1} & 0 \\ 0 & Y_{\infty,2} \end{pmatrix}. \tag{5.48}$$

Simple algebraic computation show that the only "stable" solutions to Equations 5.45–5.46, *i.e.*, solutions such the column of the matrices

$$\begin{pmatrix} I \\ X_\infty \end{pmatrix}, \quad \begin{pmatrix} I \\ Y_\infty \end{pmatrix} \tag{5.49}$$

span the $2n$-dimensional stable invariant subspace of the associated Hamiltonian matrices, are

$$X_\infty = \beta \begin{pmatrix} \Omega & 0 \\ 0 & I \end{pmatrix}, \quad Y_\infty = \beta \begin{pmatrix} \Omega^{-1} & 0 \\ 0 & I \end{pmatrix}. \tag{5.50}$$

With this selection, conditions (i) and (ii) are the same. From condition (i), the following equation must hold:

$$-2\beta\Lambda + \beta^2\left(\gamma^{-2}(\alpha\Lambda + B_{u2}B_{u2}^{\mathrm{T}}) - B_{u2}B_{u2}^{\mathrm{T}}\right) + \alpha\Lambda + B_{u2}B_{u2}^{\mathrm{T}} = 0, \tag{5.51}$$

which is equivalent to

$$\begin{aligned} -2\beta + \beta^2\gamma^{-2}\alpha + \alpha &= 0, \\ \beta^2\gamma^{-2} - \beta^2 + 1 &= 0, \end{aligned} \tag{5.52}$$

i.e.,

$$\beta = \sqrt{\frac{\gamma^2}{\gamma^2 - 1}}, \quad \alpha = 2\sqrt{1-\gamma^{-2}}. \tag{5.53}$$

Condition (iii) is satisfied if and only if $\beta^2 < \gamma^2$, which, together with Equation 5.53, implies

$$\gamma > \sqrt{2}. \tag{5.54}$$

Moreover, it is easy to show that the above condition is also necessary for the stability of the solutions of the ARE. The structure of the controller is obtained from [24]. The stability of the controller is easily checked by observing that the matrix X_∞ solves the Lyapunov equation

$$A_\infty^{\mathrm{T}} X_\infty + X_\infty A_\infty + 2\frac{\gamma^2-1}{\gamma^2-2}\sqrt{\frac{\gamma^2-1}{\gamma^2}}\begin{pmatrix} 0 & 0 \\ 0 & \Lambda_\infty\Lambda_\infty^{\mathrm{T}} \end{pmatrix} = 0, \tag{5.55}$$

and the stability of the controller follows from the stabilisability of the pair (A, B_u), which in turn implies the stabilisability of the pair $(A, [0 \quad \Lambda_\infty^{\mathrm{T}}]^{\mathrm{T}})$. By Theorem 5.1, the controller also has m zeros at the origin. The situation is more complex in the optimal case. Indeed, for $\gamma = \sqrt{2}$ all the three conditions fail, and a different approach must be used. It is known ([24], Theorem 14.6) that in the optimal case full-column rank matrices

$$\begin{pmatrix} X_{\infty 1} \\ X_{\infty 2} \end{pmatrix}, \quad \begin{pmatrix} Y_{\infty 1} \\ Y_{\infty 2} \end{pmatrix}$$

have to be sought such that

$$H_\infty \begin{pmatrix} X_{\infty 1} \\ X_{\infty 2} \end{pmatrix} = \begin{pmatrix} X_{\infty 1} \\ X_{\infty 2} \end{pmatrix} T_X, \ \mathrm{Re}\lambda_i(T_X) \le 0, \forall i, \tag{5.56}$$

$$J_\infty \begin{pmatrix} Y_{\infty 1} \\ Y_{\infty 2} \end{pmatrix} = \begin{pmatrix} Y_{\infty 1} \\ Y_{\infty 2} \end{pmatrix} T_Y, \ \mathrm{Re}\lambda_i(T_Y) \le 0, \forall i, \tag{5.57}$$

with

$$X_{\infty 1}^{\mathrm{T}} X_{\infty 2} = X_{\infty 2}^{\mathrm{T}} X_{\infty 1}, \tag{5.58}$$

$$Y_{\infty 1}^{\mathrm{T}} Y_{\infty 2} = Y_{\infty 2}^{\mathrm{T}} Y_{\infty 1}, \tag{5.59}$$

and

$$\begin{pmatrix} X_{\infty 2}^{\mathrm{T}} X_{\infty 1} & \gamma^{-1} X_{\infty 2}^{\mathrm{T}} Y_{\infty 2} \\ \gamma^{-1} Y_{\infty 2}^{\mathrm{T}} X_{\infty 1} & Y_{\infty 2}^{\mathrm{T}} Y_{\infty 1} \end{pmatrix} \geq 0, \tag{5.60}$$

where H_∞ and J_∞ are the Hamiltonian matrices

$$H_\infty = \begin{pmatrix} A & \gamma^{-2} B_w B_w^{\mathrm{T}} - B_u B_u^{\mathrm{T}} \\ -C_z^{\mathrm{T}} C_z & -A^{\mathrm{T}} \end{pmatrix}, \tag{5.61}$$

$$J_\infty = \begin{pmatrix} A^{\mathrm{T}} & \gamma^{-2} C_z^{\mathrm{T}} C_z - C_y^{\mathrm{T}} C_y \\ -B_w B_w^{\mathrm{T}} & -A \end{pmatrix}. \tag{5.62}$$

It is easy to show that the above equations are satisfied by

$$X_{\infty 1} = I,\, X_{\infty 2} = X_\infty, \tag{5.63}$$

$$Y_{\infty 1} = I,\, Y_{\infty 2} = Y_\infty, \tag{5.64}$$

where X_∞ and Y_∞ are given by Equation 5.50 and hence

$$T_X = \begin{pmatrix} 0 & I \\ -\Omega & -(1-2\gamma^{-2})\Lambda - B_{u2} B_{u2}^{\mathrm{T}} \end{pmatrix}. \tag{5.65}$$

By letting $\varepsilon = 1 - (\beta/\gamma)^2 = (\gamma^2 - 2)/(\gamma^2 - 1) > 0$, the controller is described by the *singularly perturbed* equations

$$\varepsilon \dot{x}_c = (\varepsilon T_X - B_u B_u^{\mathrm{T}}) x_c + B_u y, \tag{5.66}$$

$$u = -B_u^{\mathrm{T}} x_c. \tag{5.67}$$

Letting $\varepsilon = 0$ the *reduced-order system* is obtained and the control is $u = -K_{\mathrm{opt}} y$.

□

The meaning of the optimality of the above controller should be suitably specified. Indeed, α depends on γ, and by changing γ *different* control problems must be solved. However, an optimal controller computed by using a different approach could result in not having the zeros at the origin, thus losing the important bandpass property of the controller. On the other hand, it must be stressed that in the above theorem when varying γ from $\sqrt{2}$ to ∞, α varies on a narrow interval, *i.e.*, $[\sqrt{2}, 2]$, hence from a physical point of view this has minor consequences in most applications. The above considerations lead to the definition of optimality of the proposed controller, in terms of minimising the $\mathcal{H}_\infty$ norm while preserving m zeros at the origin and stability of the controller.

Note that the limit controller obtained in the singular case $\gamma = \sqrt{2}$, results in the static velocity feedback of Equation 5.43. However, in practical applications this would yield a too large controller bandwidth. Hence controller 5.38 is preferred in vibration control applications.

The controller in Theorem 5.2 is the *central $\mathcal{H}_\infty$ controller* [24]. Thus a free parameter Q is still available, and the complete parametrisation of all the $\mathcal{H}_\infty$ controllers is

$$\dot{\zeta} = \begin{pmatrix} A_\infty - \frac{\gamma^2-1}{\gamma^2-2} B_u D_q B_u^{\mathrm{T}} & \frac{\gamma^2-1}{\gamma^2-2} B_u C_q \\ -B_q B_u^{\mathrm{T}} & A_q \end{pmatrix} \zeta + \begin{pmatrix} \frac{\gamma^2-1}{\gamma^2-2} B_u \left(D_q + \frac{\gamma}{\sqrt{\gamma^2-1}} I \right) \\ B_q \end{pmatrix} y, \quad (5.68)$$

$$u = \left(-\left(D_q + \frac{\gamma}{\sqrt{\gamma^2-1}} I \right) B_u^{\mathrm{T}} \;\; C_q \right) \zeta + D_q y, \quad (5.69)$$

where (A_q, B_q, C_q, D_q) is a n_q-order realisation of the stable transfer matrix $Q(s)$, with $\|Q\|_\infty < \gamma$ and $\zeta \in \mathbb{R}^{2n+n_q}$ is the state of the central controller augmented with that of the parameter $Q(s)$. Once again, note that using a generic Q could result in losing stability or zeros at the origin. Nevertheless, suitable choices of Q are possible. For instance, if a bandpass transfer matrix $Q(s)$ is considered, with m zeros at the origin, then the resulting controller is again a bandpass, with m zeros at the origin, as can be easily proved by computing the DC gain. Thus a *whole family* of bandpass controllers with prescribed $\mathscr{H}_\infty$ closed loop norm is obtained.

Finally, robustness issues of the proposed solution can be analysed. In fact, it is well known that $\mathscr{H}_\infty$ controllers are able to withstand closed loop disturbances of the form $w = \Delta z$, with $\|\Delta\|_\infty < 1/\gamma$. When using the above matrices B_w and C_z, the result is a *guaranteed* stability margin for uncertainties entering in the range of the input matrix B_u and against uncertainties of the damping ratios.

5.6 $\mathscr{H}_2$ Controller Design

It is well known that an $\mathscr{H}_2$ performance index is minimised when $\gamma \to \infty$ in the $\mathscr{H}_\infty$ controller. In this case, the controller

$$K_2(s) = -B_u^{\mathrm{T}} \left(sI - A + 2B_u B_u^{\mathrm{T}} \right)^{-1} B_u \quad (5.70)$$

is obtained and $\alpha = 2$. Note that this controller is similar to the "symmetric output feedback" controller given by Junkins [15], while the cost index is similar to that proposed by Halevi [14]. Moreover, the controller implementation in Figure 5.2 shows that the optimal controller structure resembles the classic "internal model principle" controller [8], with an extra static feedback loop. In the figure, P_{22} denotes the transfer matrix from the input u to the output y, hence the controller is a "copy" of the plant with a negative feedback with gain $\alpha = 2$.

In the $\mathscr{H}_2$ case, the parametrisation of the (non-optimal) controllers yields

$$\dot{\zeta} = \begin{pmatrix} A_\infty - B_u D_q B_u^{\mathrm{T}} & B_u C_q \\ -B_q B_u^{\mathrm{T}} & A_q \end{pmatrix} \zeta + \begin{pmatrix} B_u (D_q + I) \\ B_q \end{pmatrix} y, \quad (5.71)$$

$$u = \left(-(D_q + I) B_u^{\mathrm{T}} \;\; C_q \right) \zeta + D_q y, \quad (5.72)$$

hence the symmetry of the controller is even more apparent than in the $\mathscr{H}_\infty$ case.

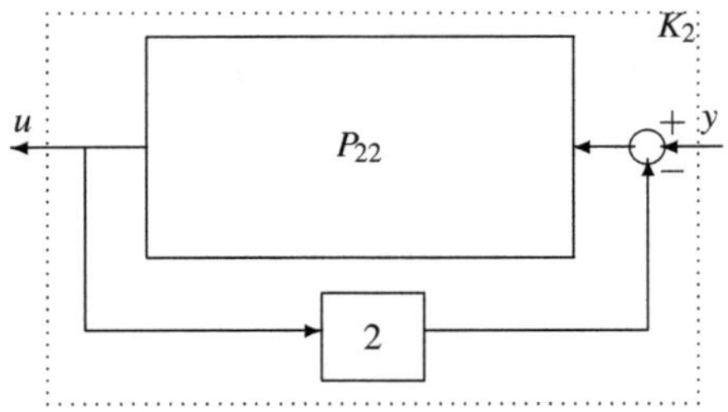

Figure 5.2 $\mathscr{H}_2$ controller

Table 5.1 Open-loop natural frequencies and damping ratios

ζ_i	ω_{n_i} [rad/s]
0.0010	129.17
0.0030	158.58
0.0020	208.69
0.0050	341.35
0.0070	354.12
0.0002	512.7
0.0080	622.49
0.0	805.51

5.7 Numerical Example

In order to test the proposed control strategy, the procedure is applied to a flexible system with $n = 8$ modes, $m = 3$ inputs and outputs, whose model (described in Equations 5.10–5.12) has the parameters (ζ_i, ω_{n_i}) in Table 5.1 and input matrix B_{u2}

$$B_{u2} = \begin{pmatrix} 3.339 & 1.033 & 2.350 \\ 1.672 & 0.132 & 1.010 \\ 0.558 & 0.798 & 0.730 \\ 7.102 & 4.688 & 7.718 \\ -2.531 & -0.813 & -2.290 \\ -4.742 & 11.670 & 5.716 \\ 0.025 & -0.019 & 0.019 \\ 2.528 & 6.278 & -17.187 \end{pmatrix}. \tag{5.73}$$

The model is not asymptotically stable but it is stabilisable, with lightly damped modes. The selection of the disturbance matrix has been made in agreement with Equation 5.14 in order to verify properties, performances and robustness of the strongly stabilising $\mathscr{H}_\infty$ controller presented in Theorem 5.2. The value of the parameter α has been fixed according to Equation 5.41. In Figure 5.3, with a value of $\gamma = 1.5$, the singular values of controller are presented, showing the bandpass property. The controller has three zeros at $s = 0$, while the remaining are listed in

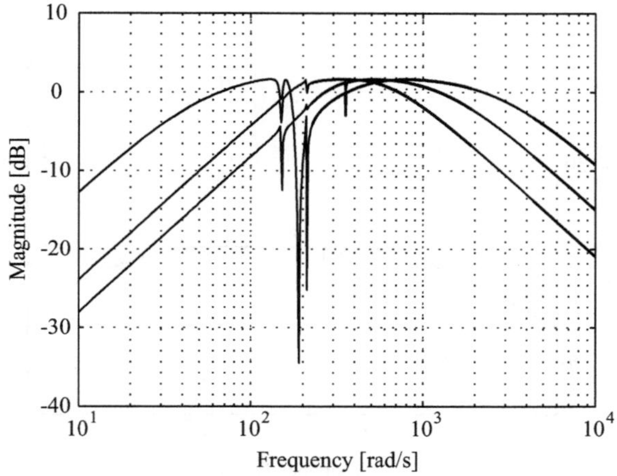

Figure 5.3 Singular values of the controller

Table 5.2 Zeros of the controller (excluding the three at $s = 0$)

ζ_i	ω_{n_i} [rad/s]
0.00026	151
0.00024	212
0.00028	189
0.00075	354
0.00089	622

Table 5.2. In Figure 5.4 the performance condition 5.42 for $\gamma = 1.5$ is verified by considering the maximum singular value of T_{zw}.

Moreover, the robustness of the $\mathscr{H}_\infty$ controller has been tested and the results are presented in Figure 5.5, where the maximum singular value of T_{zw} is reported, simulating a multiplicative uncertainty of 50% for all ω_{n_i}. Obviously, condition 5.42 is no longer verified, but closed loop stability is preserved.

Next, a $\mathscr{H}_2$ controller is designed, according to Equation 5.70. The design is much simpler, and the performance is good for vibrating systems, as shown in Figure 5.6, where the closed loop performances of the $\mathscr{H}_2$ controller are compared to open loop and to $\mathscr{H}_\infty$ controller.

Note that the reduced performances of the $\mathscr{H}_2$ controller are a consequence of the reduced gain of the controller rather than being structural to the method, as shown by the plots in Figure 5.7.

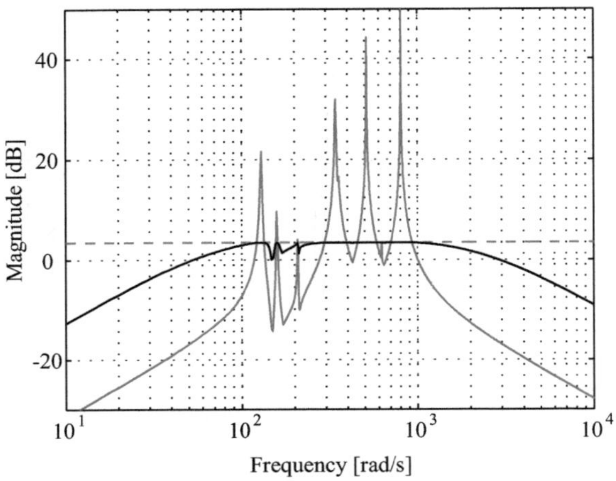

Figure 5.4 Maximum singular value of T_{zw} with $\gamma = 1.5$ (*dashed grey line*): open loop (*solid grey line*), closed loop (*solid black line*)

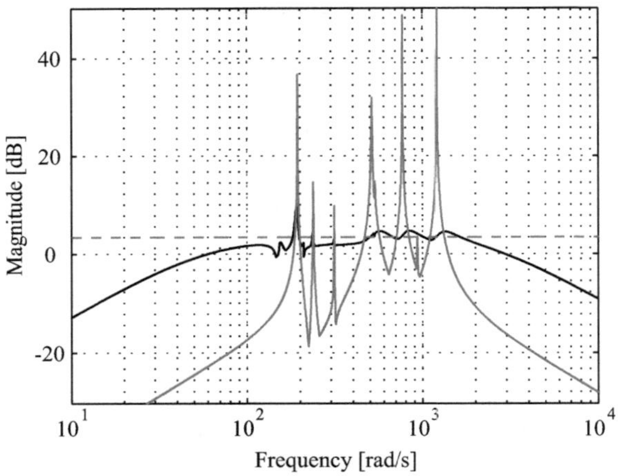

Figure 5.5 Maximum singular value of T_{zw} with $\gamma = 1.5$ (*dashed grey line*) and with a multiplicative uncertainty on ω_n of 50%: open loop (*solid grey line*) and closed loop (*solid black line*)

5.8 Further Considerations on the $\mathscr{H}_2$ Controller

The above consideration leads to the necessity to explore further the possibilities offered by the simpler $\mathscr{H}_2$ strategy.

A first issue to address is the choice of the scaling matrices, which are the only degrees of freedom left to the designer in the presented procedure. In fact, in practical applications, the scaling may have physical justification, but often it is not easy to define the exact choice of the scaling matrix. More frequently, only some global parameters, *e.g.*, the "magnitude" (*i.e.*, the norm) of the scaling matrix, can be as-

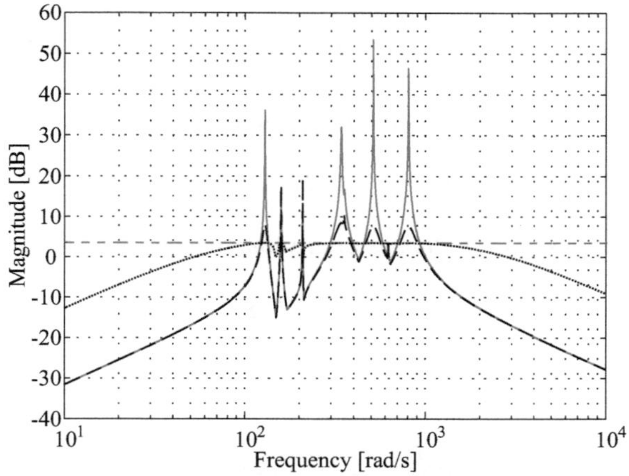

Figure 5.6 Maximum singular value of T_{zw}: open loop (*solid grey line*), closed loop $\mathcal{H}_2$ (*dashed black line*), closed loop $\mathcal{H}_\infty$ (*dotted black line*) with $\gamma = 1.5$ (*dashed grey line*)

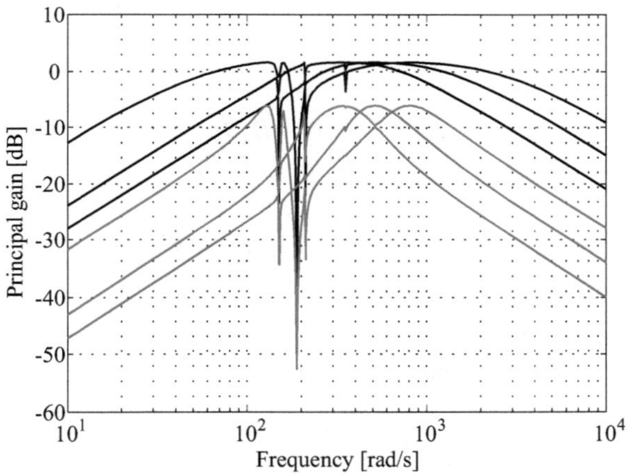

Figure 5.7 Singular values of $\mathcal{H}_2$ (*solid grey lines*) and $\mathcal{H}_\infty$ controllers (*solid black lines*)

signed. It is *not* frequently that the directions of the column vectors of the scaling matrices are also rigidly constrained so that usually only diagonal scaling matrices are chosen. There are seldom physical justifications for selecting off-diagonal matrix entries. In these cases, seeking scaling matrices resulting from the solution of a mathematically well posed optimisation problem makes sense. Moreover, from the results in the previous sections, it is known that changing the scaling matrix does not change the properties of the controller (stability, bandpass frequency shape). What changes is the performance of the closed loop system in terms of attainable value of the $\mathcal{H}_2$ norm of the LFT T_{zw}. The first step is to compute this norm, and the answer to this issue is provided by the following lemma.

Lemma 5.2. *Let the system be given by Equations 5.10–5.13 with structure in Equation 5.6 and with* $\zeta_i > 0$, $i = 1, \ldots, n$ *and consider the controller 5.70. Let* T_{zw} *be the LFT* $T_{zw} = \mathscr{F}_l(P, K)$.

Let moreover X be the controllability Gramian of the system

$$\dot{x} = (A - B_u B_u^{\mathrm{T}})x + B_u u, \tag{5.74}$$

and partition X into square n-dimensional submatrices

$$X = \begin{pmatrix} X_{11} & X_{12} \\ X_{12}^{\mathrm{T}} & X_{22} \end{pmatrix}. \tag{5.75}$$

Then the $\mathscr{H}_2$ *norm of* T_{zw} *is*

$$\|T_{zw}\|_2^2 = \mathrm{tr}(B_{w2}^{\mathrm{T}}(X_{22} + I)B_{w2}) + \mathrm{tr}(B_{u2}^{\mathrm{T}} X_{22} B_{u2}). \tag{5.76}$$

Proof. Initially, compute a space state realisation of the LFT T_{zw}

$$T_{zw} = \left(\begin{array}{c|c} A_c & B_c \\ \hline C_c & 0 \end{array}\right), \tag{5.77}$$

where

$$A_c = \begin{pmatrix} A & -B_u B_u^{\mathrm{T}} \\ B_u B_u^{\mathrm{T}} & A - 2B_u B_u^{\mathrm{T}} \end{pmatrix}, \tag{5.78}$$

$$B_c = \begin{pmatrix} 0 & 0 \\ B_{w2} & 0 \\ 0 & B_u \end{pmatrix}, \quad C_c = B_c^{\mathrm{T}}. \tag{5.79}$$

In order to compute the $\mathscr{H}_2$ norm of the closed loop system, the controllability Gramian, Y, of T_{zw} must be computed. Partition Y as

$$Y = \begin{pmatrix} Y_{11} & Y_{12} \\ Y_{12}^{\mathrm{T}} & Y_{22} \end{pmatrix}. \tag{5.80}$$

Solving the Gramian equation reduces to solving the following three linear matrix equalities:

$$AY_{11} + Y_{11}A^{\mathrm{T}} - Y_{22}B_u B_u^{\mathrm{T}} - B_u B_u^{\mathrm{T}} Y_{22} + B_w B_w^{\mathrm{T}} = 0, \tag{5.81}$$
$$AY_{12} - B_u B_u^{\mathrm{T}} Y_{22} + Y_{11} B_u B_u^{\mathrm{T}} + Y_{22}(A^{\mathrm{T}} - 2B_u B_u^{\mathrm{T}}) = 0, \tag{5.82}$$
$$(A - 2B_u B_u^{\mathrm{T}})Y_{22} + Y_{22}(A - 2B_u B_u^{\mathrm{T}})^{\mathrm{T}} + B_u B_u^{\mathrm{T}} Y_{12} + Y_{12}^{\mathrm{T}} B_u B_u^{\mathrm{T}} + B_u B_u^{\mathrm{T}} = 0. \tag{5.83}$$

Algebraic computation shows that the matrices Y_{11} and Y_{22} satisfy

$$A(Y_{11} - Y_{22}) + (Y_{11} - Y_{22})A^{\mathrm{T}} + \begin{pmatrix} 0 & 0 \\ 0 & 2\Lambda \end{pmatrix} = 0. \tag{5.84}$$

The Lyapunov equation 5.84 has the solution

$$Y_{11} - Y_{22} = \begin{pmatrix} \Omega^{-1} & 0 \\ 0 & I \end{pmatrix}. \tag{5.85}$$

This solution is also the only solution if and only if the well known condition $\lambda_i(A) + \lambda_j(A) \neq 0 \; \forall i, j$ holds, which in the present case holds true, since the open loop system is asymptotically stable by hypothesis. Moreover, it is easy to show that $Y_{12} = Y_{22}$, hence the Gramian can be expressed as

$$Y = \begin{pmatrix} Z + Y_{22} & Y_{22} \\ Y_{22} & Y_{22} \end{pmatrix}, \tag{5.86}$$

where

$$Z = \begin{pmatrix} \Omega^{-1} & 0 \\ 0 & I \end{pmatrix}. \tag{5.87}$$

Noting that the matrix $B_u B_u^{\mathrm{T}}$ is Z-invariant, from Equation 5.82 it is clear that Y_{22} solves Equation 5.81. Finally, taking into account the structure of the matrices B_u and B_w in B_c, Equation 5.76 is readily proved. □

The above Lemma shows that in order to compute the $\mathscr{H}_2$ norm of the $4n$-order closed loop system, only an $(n \times n)$ matrix is needed. However, computing this matrix in general requires solving a Lyapunov equation of order $2n$. In some cases closed loop solutions can be found, as in the case presented in the following corollary.

Corollary 5.1. *Assume that system 5.10 has as many input as modes and let the input matrix to have full-column rank, so that without loss of generality it can be assumed that $B_{u2} = I$. Then the controller $K_2(s)$ in Equation 5.70 yields a closed loop transfer function with $\mathscr{H}_2$ norm*

$$\|T_{zw}\|_2^2 = 2\left(n + 2\sum_{i=1}^{n} \zeta_i \omega_{n_i}\right). \tag{5.88}$$

Proof. The corollary easily follows from simple algebraic manipulations of the result of Lemma 5.2. □

Although Corollary 5.1 has limited practical applicability, such as assuming as many input as modes is highly unrealistic when dealing with systems having a theoretically infinite number of modes, it is useful in determining a lower bound for the $\mathscr{H}_2$ norm of the closed loop system. In fact, it is clear that in the situation considered in the corollary all the modes are controlled independently of each other, thus bringing the closed loop norm to its smallest value. Note that in this case, the $\mathscr{H}_2$ norm depends on the system complexity (the number of modes) and on the product of damping and frequency of the modes, thus *high frequency* modes affect the $\mathscr{H}_2$ norm more heavily (which is an expected result, based on physical considerations) and *low damping* reduces the norm, which is the main difference with the $\mathscr{H}_\infty$ norm. This is, moreover, a further justification of the selection of matrix B_{w2} in Equation 5.14.

The results in this section are the basis for the choice of the scaling matrix, which will be dealt with in the next section.

5.9 Optimal Scaling Matrix Selection

In this section the hypothesis of scaled input and outputs is removed. This means that the input matrix in Equation 5.10 must be written as

$$B_u = \tilde{B}_u R, \tag{5.89}$$

where $\tilde{B}_u$ is the true, physical control input matrix and R is a scaling matrix. Analogously, the output matrix will be $C_y = R^{\mathrm{T}} \tilde{C}_y$ and the resulting controller $\tilde{K}_2$ will multiply both matrices on the input and output

$$\tilde{K}_2(s) = R K_2(s) R^{\mathrm{T}}, \tag{5.90}$$

as is well known from the scaling procedure. Keeping the symmetry of the input and output matrix is needed for retaining the properties of the controller, hence only one matrix needs to be selected, and this is the subject of the next section.

5.9.1 *Formulation of the Optimsation Problem*

Before detailing the selection of the scaling matrix, a preliminary remark is in order. Indeed, aiming at the direct minimisation of the $\mathscr{H}_2$ norm of the closed loop system, *e.g.*, by using Lemma 5.2, is not correct. In fact, since the matrices B_w and C_z depend on the scaling matrix R themselves, simply comparing different values of $||T_{zw}||_2$ obtained with different values of R would lead to unfair comparisons, as also the weighting matrices (and the transfer function T_{zw}) change. This produces "odd" results, as is the case if only the gain of the matrix R is increased: the value of $||T_{zw}||_2$ does not change, but, as the controller has higher bandwidth and gain, the controller authority increases, and the vibration reduction is clearly higher. A more interesting and useful approach is to employ an index sensitive to the performance improvement when closing the control loop. A suitable way to accomplish this task is to define an index able to measure the "potential for improvement", as the ratio between open loop and closed loop performances, as follows:

$$J(R) = \frac{||T_{zw}^0||_2}{||T_{zw}||_2}, \tag{5.91}$$

where T_{zw}^0 is the open loop performance, *i.e.*, the transfer function between z and w when $u = 0$. Moreover, as the properties of the controller, including stability and stabilisation capability, are independent of the norm of the matrix R, a suitable optimisation strategy is to compute a normalised scaling matrix first and then to adjust the gain (*i.e.*, the norm of R) so as to achieve a required vibration attenuation level (of course taking into account the control energy). The problem, then, can be formulated as the optimisation problem

$$J_{\text{opt}} = \sup_{\|R\|=1} \frac{\|T_{zw}^0\|_2}{\|T_{zw}\|_2}. \tag{5.92}$$

Based on the results of Lemma 5.2 and on standard results of the computation of $\mathscr{H}_2$ norm, the index $J(R)$ can be computed as follows. First, compute X_0, the controllability Gramian of the open loop system by solving

$$AX_0 + X_0A^{\mathrm{T}} + B_wB_w^{\mathrm{T}} = 0. \tag{5.93}$$

Next compute

$$\|T_{zw}^0\|_2^2 = \mathrm{tr}(B_{w2}^{\mathrm{T}}X_{02}B_{w2}), \tag{5.94}$$

where X_{02} is the lower right block (the $(2,2)$-block) of the matrix X_0, partitioned accordingly to the partition of B_w. Analogously, compute $\|T_{zw}\|_2$ by using the result of Lemma 5.2. Note that computing the cost index 5.92 amounts to solving a nonlinear optimisation problem, as both the Gramians and the weighting matrices depend on R. A simpler approach is to adopt a two-stage procedure, computing the Gramians first and next solving the optimisation problem

$$\sup_{\|R\|=1} \frac{\mathrm{tr}(B_{w2}^{\mathrm{T}}X_{02}B_{w2})}{\mathrm{tr}(B_{w2}^{\mathrm{T}}(X_{22}+I)B_{w2}) + \mathrm{tr}(B_{u2}^{\mathrm{T}}X_{22}B_{u2})}, \tag{5.95}$$

assuming X_{02} and X_{22} constant. Then, based on the new value of R, again the Gramians are evaluated and the procedure is repeated until the optimal value is reached. The issue of the convergence of the algorithm, although very important, will not be discussed here. The following theorem solves problem 5.95.

Theorem 5.3. *Consider the optimisation problem 5.95, where X_{02} and X_{22} are constant matrices. Let $S = \mathrm{blockdiag}_m(\tilde{B}\mathrm{T}_{u2}X_{02}\tilde{B}_{u2})$ and $T = \mathrm{blockdiag}_m(\tilde{B}\mathrm{T}_{u2}(2X_{22} + I)\tilde{B}_{u2})$, where for any square matrix H, $\mathrm{blockdiag}_m(H)$ denotes the block diagonal matrix with the block H repeated m times. Moreover, let $v_{\max}$ be the eigenvector corresponding to the largest eigenvalue $\lambda_{\max}$ of the generalised eigenvalue problem*

$$(S+\sigma I)v = \lambda(T+\tau I)v, \tag{5.96}$$

where $\sigma = 2\mathrm{tr}(\Lambda X_{02})$ and $\tau = 2\mathrm{tr}\left[\Lambda(X_{22}+I)\right]$. Then the solution to problem 5.95 is

$$\mathrm{vec}(R) = v_{\max}, \tag{5.97}$$

where $\mathrm{vec}(R)$ denotes the vector obtained by stacking the columns of the matrix R.

Proof. First, note that the numerator of the objective function in problem 5.95 can be rewritten as

$$\mathrm{tr}(B_{w2}^{\mathrm{T}} X_{02} B_{w2}) = \mathrm{tr}(B_{w2} B_{w2}^{\mathrm{T}} X_{02}) = \mathrm{tr}\left((2\Lambda + B_{u2} B_{u2}^{\mathrm{T}}) X_{02}\right) =$$
$$2\mathrm{tr}(\Lambda X_{02}) + \mathrm{tr}(B_{u2} B_{u2}^{\mathrm{T}} X_{02}) = 2\mathrm{tr}(\Lambda X_{02}) + \mathrm{tr}(R^{\mathrm{T}} \tilde{B}\mathrm{T}_{u2} X_{02} \tilde{B}_{u2} R). \tag{5.98}$$

Analogously, the denominator can be written as

$$\mathrm{tr}(B_{w2}^{\mathrm{T}}(X_{22}+I)B_{w2}) + \mathrm{tr}(B_{u2}^{\mathrm{T}} X_{22} B_{u2}) =$$
$$2\sum_{i=1}^{n} \lambda_i + 2\mathrm{tr}(\Lambda X_{22}) + \mathrm{tr}(R^{\mathrm{T}} \tilde{B}\mathrm{T}_{u2}(2X_{22}+I)\tilde{B}_{u2} R). \tag{5.99}$$

By noting that, for an $r \times r$ matrix F, the following holds:

$$\mathrm{tr}(F) = \begin{pmatrix} e_1^{\mathrm{T}} & \cdots & e_r^{\mathrm{T}} \end{pmatrix} \mathrm{blockdiag}_r(F) \begin{pmatrix} e_1 \\ \vdots \\ e_r \end{pmatrix}, \tag{5.100}$$

where e_i is the ith vector of the unit basis of $\mathbb{R}^r$, *i.e.*, the vector with 1 at entry i and zero elsewhere and noting also that

$$\mathrm{blockdiag}_m(R) \begin{pmatrix} e_1 \\ \vdots \\ e_m \end{pmatrix} = \mathrm{vec}(R). \tag{5.101}$$

The index J in Equation 5.91 can be written as

$$\frac{(\mathrm{vec}(R))^{\mathrm{T}} S \mathrm{vec}(R) + \sigma I}{(\mathrm{vec}(R))^{\mathrm{T}} T \mathrm{vec}(R) + \tau I}. \tag{5.102}$$

Finally, it is not hard to prove that minimising the term 5.102 under the norm constraint is equivalent to solving the generalised eigenvalue problem 5.96. □

The computation in Theorem 5.3 can be simplified by using the following corollary.

Corollary 5.2. *Consider the optimisation problem 5.95, where* X_{02} *and* X_{22} *are constant matrices. Let* $S_0 = \tilde{B}\mathrm{T}_{u2} X_{02} \tilde{B}_{u2}$ *and* $T_0 = \tilde{B}\mathrm{T}_{u2}(2X_{22}+I)\tilde{B}_{u2}$*. Moreover, let* $v_{0\max}$ *be the eigenvector corresponding to the largest eigenvalue* $\lambda_{0\max}$ *of the generalised eigenvalue problem*

$$(S_0 + \sigma I)v = \lambda (T_0 + \tau I)v, \tag{5.103}$$

where $\sigma = 2\mathrm{tr}(\Lambda X_{02})$ *and* $\tau = 2\mathrm{tr}\left[\Lambda (X_{22}+I)\right]$*. Then the solution to problem 5.95 is*

$$R = 1_m \otimes v_{0,\max}, \tag{5.104}$$

where $1_m = \begin{pmatrix} 1 & 1 & \ldots & 1 \end{pmatrix}$, i.e., *the row vector with entry* 1 *repeated m times, and* $\otimes$ *denotes the Kronecker product.*

Proof. This follows easily noting that the solution $v_{\max}$ to the eigenvalue problem involving only block diagonal matrices 5.96 is obtained by stacking m times the eigenvector $v_{0\max}$. □

Based on the above considerations, it is possible to formulate the following algorithm to compute a suboptimal value of the index J in Equation 5.91:

1. Initialisation. Start with any matrix R (*e.g.*, $R = I$).
2. Compute the matrices B_u and B_w and the matrices X_0 and X_2 from the controllability Gramians.
3. Update the matrix R by using Theorem 5.3 and Corollary 5.2.
4. If the difference between updated and previous scaling matrices is smaller than a prescribed tolerance, stop. Otherwise, go to step 2.

The procedure will be illustrated with a numerical example in the next section.

5.9.2 Scaling Matrix Selection Example

The strategy presented in the previous section has been applied to the example presented in Section 5.7. However, since the $\mathcal{H}_2$ norm is finite only for asymptotically stable systems, some damping ($\zeta_8 = 0.005$) has been considered for the eighth mode at frequency 805.51 rad/s.

Using the above strategy, the scaling matrix

$$R_{\text{opt}} = \begin{pmatrix} 0.2042 & 0.2042 & 0.2042 \\ -0.4891 & -0.4891 & -0.4891 \\ -0.2290 & -0.2290 & -0.2290 \end{pmatrix} \tag{5.105}$$

solving problem 5.92 is obtained. Note that the columns of the matrix are equal, and the controller will be of rank 1. This is apparently a drawback of the proposed strategy, although the $\mathcal{H}_2$ index is minimised. In order to check the effectiveness of the selection of the scaling matrix a comparison with the case $R = I$ has been made. Moreover, to emphasise the effect of the selection, both matrices have been multiplied by a factor 10 (remember that the strategy only selects the directions of the columns of the scaling matrix, the norm is selected independently). Thus, the comparison is between the case $R = 10R_{\text{opt}}$ and $R = 10I$.

Selecting $R = 10R_{\text{opt}}$ the value of the index $J(10R_{\text{opt}}) = 13.47$ is obtained. The open loop and closed loop singular values of the performance transfer functions are shown in Figure 5.8.

Next, the case $R = 10I$ is considered. First, the index $J(10I) = 8.34$ is computed, thus showing the good improvement obtained with the optimal strategy.[1] In Figure 5.9 the open loop and closed loop performances are compared for the case $R = 10I$.

---

[1] It has to be stressed that this is not trivial, since the optimality has been obtained in the case $\|R\| = 1$.

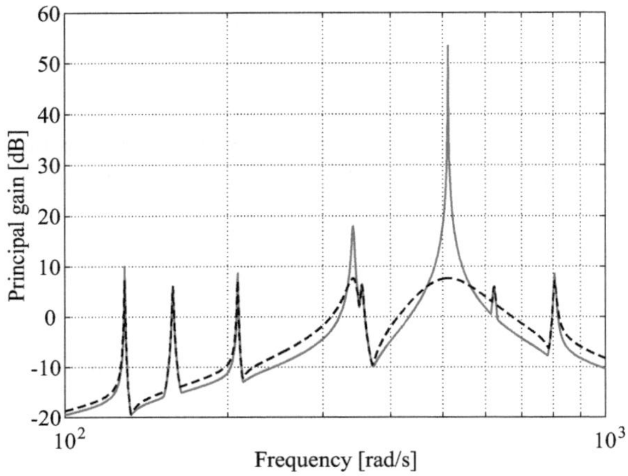

Figure 5.8 Open loop (*solid grey line*) and closed loop (*dashed black line*) principal gains (optimal R)

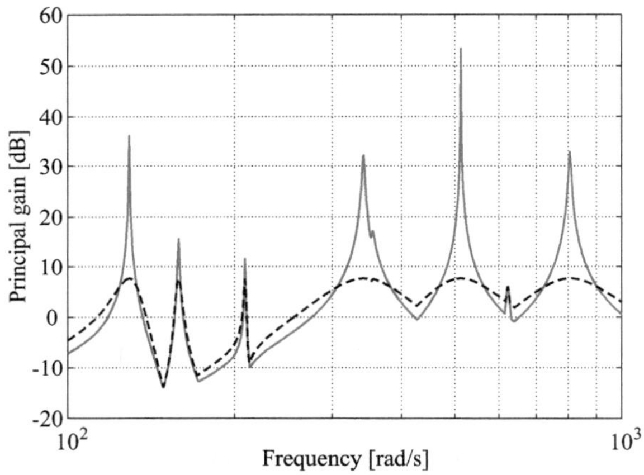

Figure 5.9 Open loop (*solid grey line*) and closed loop (*dashed black line*) principal gains ($R = 10I$)

The largest principal gain of the two controllers is shown in Figure 5.10. It is apparent that the optimal controller is more selective on the frequency range and has a narrower bandwidth, while in the frequency range $[430, 640]$ rad/s (roughly) the two controller gains coincide.

A final consideration concerns comparison of the performances. As shown in Figures 5.8 and 5.9, both open loop and closed loop performance vary when changing R. Thus, a fair comparison is obtained when comparing the two cases between the control inputs and measured outputs, bypassing weighting matrices. The result of comparison is shown in Figure 5.11.

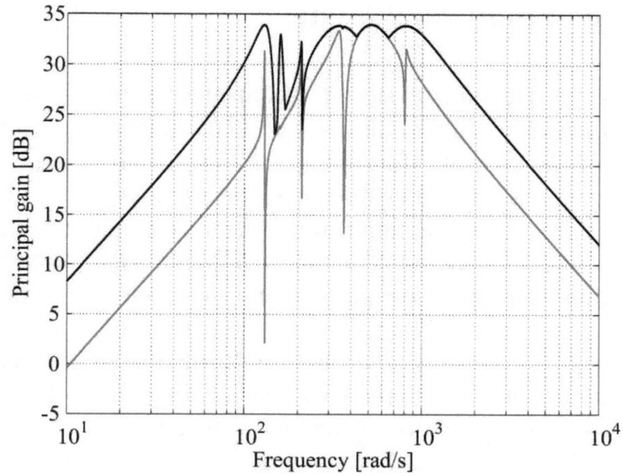

Figure 5.10 Singular values of $\mathscr{H}_2$ controllers: $R = 10I$ (*solid black line*), optimal R (*solid grey line*)

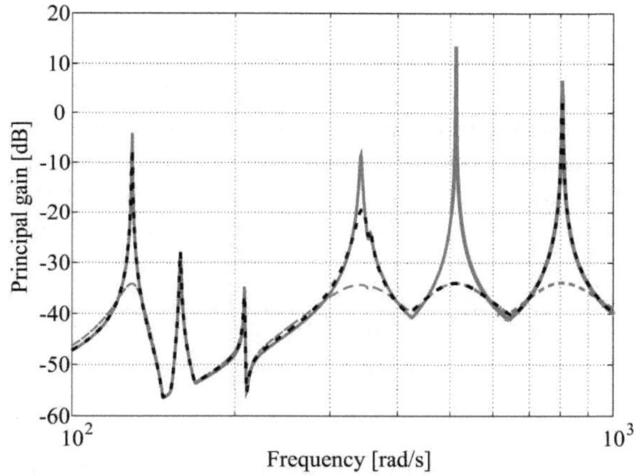

Figure 5.11 Comparison of performances: open loop (*solid grey line*), closed loop with optimal R (*dashed black line*), closed loop with $R = 10I$ (*dashed grey line*)

A comment on Figure 5.11 is in order. The optimal controller focusses its effort only on the sixth mode, the only one in the $[430, 640]\,\mathrm{rad/s}$ range, and gives some contribution on the fourth mode, while leaving the others almost unaffected. On the contrary, a non-optimal controller tries to increase the damping of *all* the modes. Thus optimality seems related to the capability to act selectively on the modes. The strategy presented in the next section will concentrate specifically on this point.

Moreover, it seems obvious that the controller effort must be focussed on the modes that are more controllable and more observable. The latter two subjects form the basis for the considerations that lead to the material developed in the next section.

5.10 Selection of Individual Modes

The objective of this section is to design a controller extending the capabilities of those previously presented in two directions: (1) the selection of individual modes on which to focus the control effort and (2) the use of non-square controllers for systems where more measurements than actuators are available. The technological motivation for the second assumption is that often it is simpler to add extra sensors, even at locations where installing actuators is difficult. The question that arises at this point is whether adding sensors permits improvements in the controlled system performance.

In order to design a stabilising controller for the flexible system refer again to the TITO standard control problem framework depicted in Figure 5.1. Also in this case, the model of the flexible system is augmented with suitable fictitious "disturbance" and "performance" matrices, thus leading to the following generalised plant P:

$$\dot{x} = Ax + B_w w + B_u u, \tag{5.106}$$

$$z = C_z x \qquad\qquad + D_{zu} u, \tag{5.107}$$

$$y = C_y x + D_{yw} w, \tag{5.108}$$

where $x = (x_1^{\mathrm{T}} \quad x_2^{\mathrm{T}})^{\mathrm{T}} \in \mathbb{R}^{2n}$ is the state vector, $w \in \mathbb{R}^{m+n_d+n_1}$ a fictitious disturbance, $u \in \mathbb{R}^m$ the control input, $z \in \mathbb{R}^{m+n_2}$ a fictitious performance output and $y \in \mathbb{R}^{m+n_d}$ the measured (control) output. The system is again assumed to be in modal form, with Equations 5.6 and 5.7 still holding. The matrices B_w and C_z are $2n \times (m+n_d+n_1)$ and $(m+n_2) \times 2n$ real matrices, respectively, with the following structure:

$$B_w = \begin{pmatrix} 0_{n\times n_1} & 0_{n\times(n_d+m)} \\ W^{1/2} & 0_{n\times(n_d+m)} \end{pmatrix}, \tag{5.109}$$

$$C_1 = \begin{pmatrix} 0_{n_2\times n} & Q^{1/2} \\ 0_{m\times n} & 0_{m\times n} \end{pmatrix}, \tag{5.110}$$

where the matrices W and Q are to be selected. Finally, for the sake of simplicity, the matrices D_{zu} and D_{yw} are chosen so as to satisfy the following conditions:

$$D_{zu}^{\mathrm{T}} C_z = 0, \quad B_w D_{yw}^{\mathrm{T}} = 0. \tag{5.111}$$

A possible selection is

$$D_{zu} = \begin{pmatrix} 0_{n_2\times m} \\ D_{zu_2} \end{pmatrix}, \quad D_{yw} = \begin{pmatrix} 0_{n_d\times n_1} & D_{yw_1} \\ 0_{m\times n_1} & D_{yw_2} \end{pmatrix}, \tag{5.112}$$

where

$$R_1 = D_{zu}^{\mathrm{T}} D_{zu} = D_{zu_2}^{\mathrm{T}} D_{zu_2} > 0, \tag{5.113}$$

$$R_2 = \begin{pmatrix} D_{yw_1} D_{yw_1}^{\mathrm{T}} & 0_{n_d\times m} \\ 0_{m\times n_d} & D_{yw_2} D_{yw_2}^{\mathrm{T}} \end{pmatrix} = \begin{pmatrix} R_{2_1} & 0 \\ 0 & R_{2_2} \end{pmatrix} > 0, \tag{5.114}$$

and R_i are weighting matrices to be selected. They are a generalisation of the scaling matrices discussed in Section 5.9.1.

As in the previous sections, also in this section, the roadmap to the controller design is the following: first, a class of bandpass controllers is characterised, next a class of stabilising controllers is defined within the set of controllers belonging to the previous class and finally, one (or more) stable controller in the last class is sought.

Starting with the class of controllers guaranteeing bandpass frequency shape, the following lemma is useful for a state space characterisation. The lemma extends the class defined in Theorem 5.1

Lemma 5.3. *Let $\mathscr{B}$ be a LTI 2n-(McMillan) degree system with minimal space state representation*

$$\mathscr{B} = C(sI - A)^{-1}B, \tag{5.115}$$

where

$$A = \begin{pmatrix} 0 & I \\ A_1 & A_2 \end{pmatrix}, \quad B = \begin{pmatrix} 0 \\ B_2 \end{pmatrix}, \quad C = \begin{pmatrix} 0 & C_2 \end{pmatrix}, \tag{5.116}$$

where all the matrices are partitioned assuming the state has been partitioned into two n-dimensional substates and the square $n \times n$ matrix A_1 is invertible. Then the system $\mathscr{B}$ is bandpass in the sense that

$$\lim_{s \to 0} \mathscr{B}(s) = \lim_{s \to \infty} \mathscr{B}(s) = 0. \tag{5.117}$$

Proof. Although the lemma can be proved as an extension of Theorem 5.1, an alternative (and simpler) proof is obtained by observing that

$$\lim_{s \to 0}(sI - A)^{-1} = \begin{pmatrix} A_1^{-1}A_2 & A_1^{-1} \\ -I & 0 \end{pmatrix}. \tag{5.118}$$

Thus, direct computation shows that the static gain is zero. The high frequency behaviour is trivial. □

The choice of a stabilising controller with the structure in Lemma 5.3 can be carried out by exploiting the following theorem.

Theorem 5.4. *Any system of the form of Equation 5.106 with matrices given by Equations 5.6, 5.7 and 5.112 is closed-loop stabilised by a controller*

$$K(s) = C_c(sI - A_c)^{-1}B_c, \tag{5.119}$$

where

$$B_c = \begin{pmatrix} 0_{m\times n} & 0_{m\times n} \\ Y_2 B_{d2} R_{2_1}^{-1} & Y_2 B_{u2} R_{2_2}^{-1} \end{pmatrix}, \tag{5.120}$$

$$C_c = -R_1^{-1}\begin{pmatrix} 0_{m\times n} & B_{u2}^{\mathrm{T}} X_2 \end{pmatrix}, \tag{5.121}$$

$$A_c = A + B_2 C_c - B_c C_2, \tag{5.122}$$

and X_2 and Y_2 are $n \times n$ diagonal matrices with non-negative entries.

Proof. Define two symmetric matrices Q and W by

$$Q = 2X_2\Lambda + X_2B_{u2}R_1^{-1}B_{u2}^{\mathrm{T}}X_2 \tag{5.123}$$

and

$$W = 2Y_2\Lambda + Y_2(B_{d2}R_{2_1}^{-1}B_{d2}^{\mathrm{T}} + B_{u2}R_{2_2}^{-1}B_{u2}^{\mathrm{T}})Y_2. \tag{5.124}$$

It is clear that both Q and W are positive semidefinite for any non-negative matrices X_2 and Y_2. Moreover, by defining the matrices

$$X = \begin{pmatrix} \Omega X_2 & 0 \\ 0 & X_2 \end{pmatrix}, \tag{5.125}$$

$$Y = \begin{pmatrix} \Omega^{-1}Y_2 & 0 \\ 0 & Y_2 \end{pmatrix}, \tag{5.126}$$

it is easy to show that $X = \mathrm{Ric}(H_2)$ and $Y = \mathrm{Ric}(J_2)$, where for a given Hamiltonian matrix Z, $\mathrm{Ric}(Z)$ denotes the *stabilising solution* of the Riccati equation associated to the Hamiltonian matrix [24], and the Hamiltonian matrices H_2 and J_2 are

$$H_2 = \begin{pmatrix} A & -B_2R_1^{-1}B_2^{\mathrm{T}} \\ -\hat{Q} & -A^{\mathrm{T}} \end{pmatrix} \tag{5.127}$$

and

$$J_2 = \begin{pmatrix} A^{\mathrm{T}} & -C_2^{\mathrm{T}}R_2^{-1}C_2 \\ -\hat{W} & -A \end{pmatrix}. \tag{5.128}$$

Finally, the matrices $\hat{Q}$ and $\hat{W}$ are

$$\hat{Q} = \begin{pmatrix} 0 & 0 \\ 0 & Q \end{pmatrix}, \tag{5.129}$$

$$\hat{W} = \begin{pmatrix} 0 & 0 \\ 0 & W \end{pmatrix}. \tag{5.130}$$

Thus, for any matrices X_2 and Y_2 the controller 5.119–5.122 solves an $\mathscr{H}_2$ problem and is therefore stabilising. □

The final issue is to guarantee the stability of the controller. Preliminarily, define

$$\hat{B}_X = B_{u2}R_1^{-1}B_{u2}^{\mathrm{T}}, \tag{5.131}$$

$$\hat{B}_Y = B_{d2}R_{2_1}^{-1}B_{d2}^{\mathrm{T}} + B_{u2}R_{2_2}^{-1}B_{u2}^{\mathrm{T}}. \tag{5.132}$$

The above theorem says that by selecting diagonal positive matrices X_2 and Y_2 the closed loop stability is assured. Hence, it is sufficient to select positive, diagonal matrices such that the controller 5.119–5.122 is stable. This can be done by solving an LMI problem, as shown by the following theorem.

Theorem 5.5. *Consider the system 5.6 and 5.7 and assume that X_2 and Y_2 are two diagonal matrices solving the LMIs*

$$X_2\hat{B}_X + \hat{B}_X X_2 + Y_2\hat{B}_Y + \hat{B}_Y Y_2 + 2\Lambda > 0, \tag{5.133}$$
$$X_2 > 0, \tag{5.134}$$
$$Y_2 > 0. \tag{5.135}$$

Then the controller K defined in Equations 5.119–5.122 strongly stabilises the system 5.6 and 5.7 and is bandpass.

Proof. The theorem can be proved by noting that the dynamic matrix of the controller is

$$A_c = \begin{pmatrix} 0 & I \\ -\Omega & -\Lambda - \hat{B}_X X_2 - Y_2\hat{B}_Y \end{pmatrix}, \tag{5.136}$$

which is stable due to Equation 5.133, as can be easily proved by using as a Lyapunov function

$$V(x) = x^{\mathrm{T}} \begin{pmatrix} \Omega^{-1} & 0 \\ 0 & I \end{pmatrix} x. \tag{5.137}$$

The rest of the theorem follows from Theorem 5.4 and Lemma 5.3. □

Note that the LMIs to be solved have a simple structure; in fact, since X_2 and Y_2 are diagonal, conditions 5.134 and 5.135 are trivial, while any scalar matrices $X_2 = \alpha I$, $Y_2 = \beta I$, with $\alpha > 0$, $\beta > 0$ are a feasible solution of inequality 5.133, thus generalising Equation 5.50.

Thus, according to the above theorem, only two positive diagonal matrices need to be selected in such a way to fulfil Equation 5.133 and a stable, stabilising, band-limited controller is easily obtained. However, this methodology says nothing about the performance of the controller. Following the proof of Theorem 5.4, the classic LQG weights can be computed *a posteriori*, thus explaining what the controller is really doing. By using expressions 5.129, 5.123, 5.130 and 5.124 it is clear that two terms are responsible for the closed loop behaviour: the diagonal matrix 2Λ and two full symmetric matrices, $\hat{B}_X$ and $\hat{B}_Y$, which do not have any peculiar structure. Thus, the diagonal matrices X_2 and Y_2 can be seen as weighting matrices for the modes of the system. This consideration leads to a control strategy aimed at reducing the contribution of single modes by simply looking for solutions of the LMIs with higher magnitude on the diagonal element corresponding to the selected mode. In other words, to reject or at least reduce the energy contribution of mode ith on the structure, it is sufficient to seek solutions with the largest ith diagonal elements.

The above considerations can be summarised in a procedure for controller design, namely:

1. Preliminarily, collect the diagonal elements of X_2 and Y_2 in the vector $\xi = (x_{2_{11}}, \ldots, x_{2_{nn}}, y_{2_{11}}, \ldots, y_{2_{nn}})^{\mathrm{T}}$, define the modes to be rejected and define a vector $c \in \mathbb{R}^n$ such that $c_j = 1$ if mode j must be rejected, $c_j = 0$ otherwise.
2. Solve the optimisation problem

$$\max \left(c^{\mathrm{T}} \quad c^{\mathrm{T}}\right) \xi, \tag{5.138}$$

subject to

$$X_2(\xi)\tilde{B}_X + \tilde{B}_X X_2(\xi) + Y_2(\xi)\tilde{B}_Y + \tilde{B}_Y Y_2(\xi) + 2\Lambda > 0, \tag{5.139}$$

$$X_2(\xi) > 0, \tag{5.140}$$

$$Y_2(\xi) > 0, \tag{5.141}$$

$$1_n^{\mathrm{T}} X_2(\xi) 1_n + 1_n^{\mathrm{T}} Y_2(\xi) 1_n < \delta, \tag{5.142}$$

where $\delta > 0$.

3. Compute the controller according to Equations 5.119–5.122.

In the above procedure, the scalar $\delta > 0$ is chosen to limit the controller gain, since Equation 5.142 can be written as

$$\mathrm{tr}(X_2) + \mathrm{tr}(Y_2) < \delta. \tag{5.143}$$

Obviously, in the above algorithm c_j can assume any real value if "partial rejection" of modes is sought. For instance, choosing $c_1 = 10$, $c_2 = 1$ and $c_3 = 0.1$ would pose different penalties on different modes, thus focussing on rejection of mode one first, next of mode two and finally of mode three. Another aspect of the presented control strategy to highlight is the selection of the matrices R_1 and, even more importantly, R_2 in Equations 5.113 and 5.114, respectively. The former is responsible for the authority of the control. The latter for the relative weights of colocated measurements with respect to non-colocated ones. By varying the magnitude of R_{2_1} relative to R_{2_2} the case of non-colocated feedback can be considered with stability guaranteed, as will be shown in the next section.

5.10.1 Mode Selection and Non-colocated Feedback Example

In order to test the effectiveness of the proposed procedure, consider again the example in Section 5.7. Since it is apparent that the main contribution to vibration is given by the sixth mode, consider the design of a controller with $c_6 = 1$, with the remaining c_i set to zero. For the sake of simplicity only one input and two outputs are considered. The actuator is colocated with the second sensor. Letting $\delta = 100$, the following situations are considered:

- colocated and non-colocated measurements are considered;
- only a colocated measurement is used by the controller;
- only a non-colocated measurement is used by the controller.

In the first case, the following weighting matrices (scalars, in this particular case) are used

$$R_1 = 10^{-1}, \quad R_{2_1} = 10^{-1}, \qquad R_{2_2} = 10^{-1}. \tag{5.144}$$

The results are given in Figures 5.12 and 5.13, and show that the sixth mode has been dramatically damped. Moreover, in order to gain insight on the effect of the

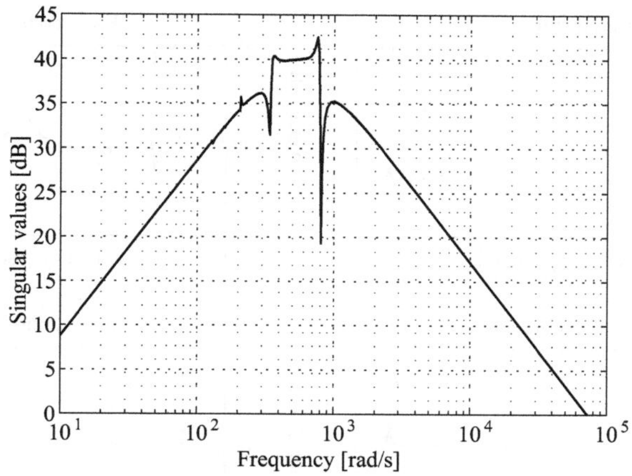

Figure 5.12 Principal gains of the controller

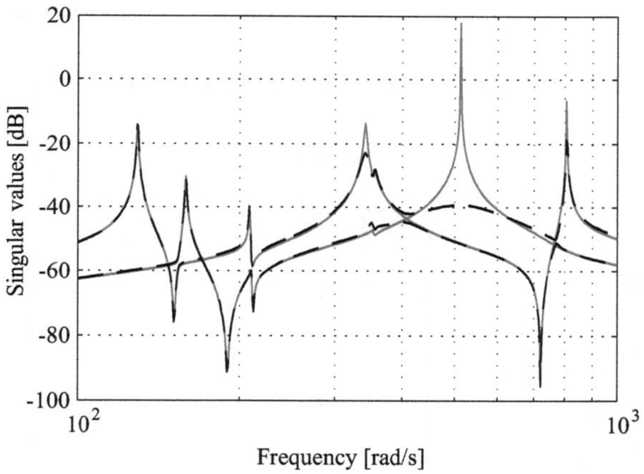

Figure 5.13 Closed loop (*dashed black lines*) and open loop (*solid grey lines*) performance

controller, disturbances have been considered entering the location where the first group of sensors are, *i.e.*, those non-colocated with actuators, thus leading to a two-input-two-output closed loop system. In Figure 5.14 the effect of the control strategy on the different channels is shown.

The design in the second and third cases produces results that are only approximate: in fact, theoretically a contribution from both sensors is still required. Nevertheless, by increasing the weighting matrices R_{2_i} it is possible to approximate the behaviour of controllers feeding back only some of the measurements. First the case of perfect colocation is considered by letting $R_{2_1} = 10^6$. The results are shown in Figures 5.15–5.17.

Next, the non-colocated case is checked, with $R_{2_2} = 10^6$. The results are shown in Figures 5.18–5.20.

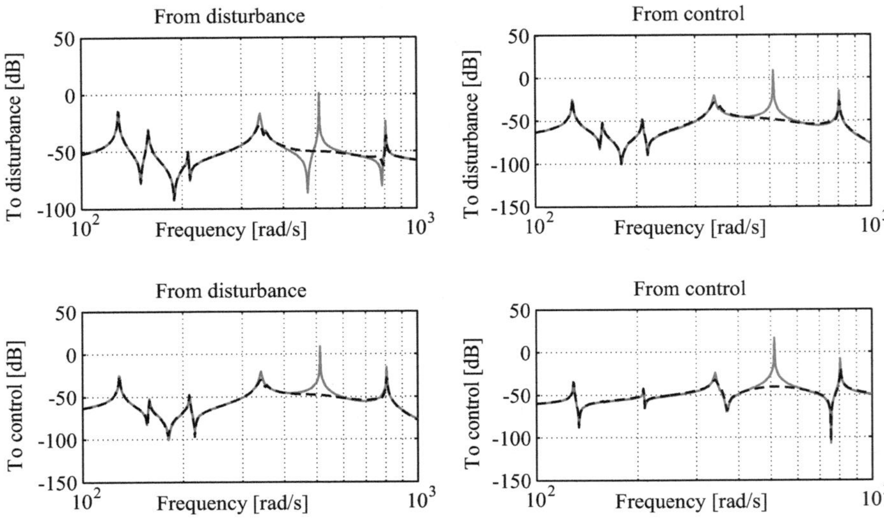

Figure 5.14 Closed loop (*dashed black lines*) and open loop (*solid grey lines*) performance for different channels

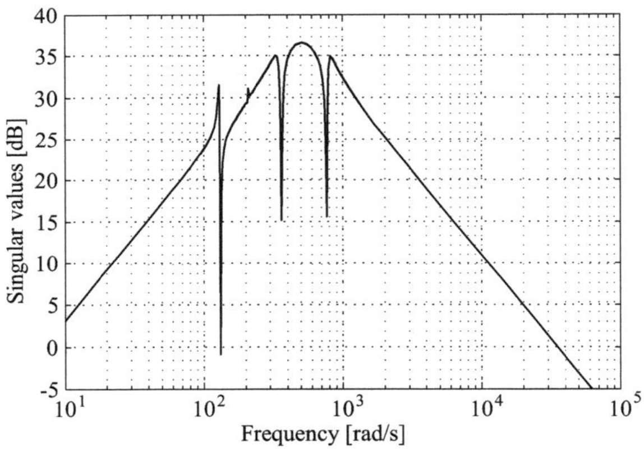

Figure 5.15 Controller principal gains, colocated case

It is worth mentioning that, although approximate, the proposed procedure is very useful from the application point of view, as the approximation is quite reasonable: for instance, in the example just presented, in the third, non-colocated case, only two controllers are theoretically needed, but computing their norms, it results that the first controller, the one between the uncolocated actuator-sensor pair has infinity norm 138, while for the colocated pair the norm is only $3.43 \cdot 10^{-5}$. Thus, it is more than reasonable to implement the controller only between the uncolocated channels.

A final consideration is in order, to emphasise the limits of the proposed approach. As in any optimisation problem, the obtained performances depend on the

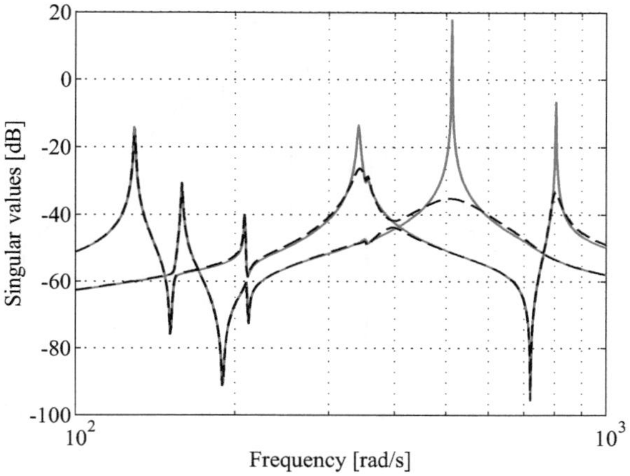

Figure 5.16 Closed loop (*dashed black lines*) and open loop (*solid grey lines*) performance, colocated case

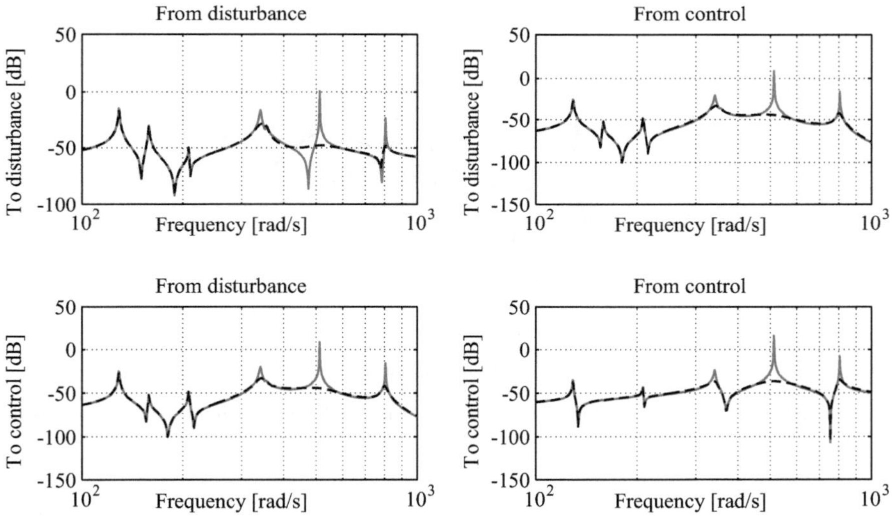

Figure 5.17 Closed loop (*dashed black lines*) and open loop (*solid grey lines*) performance for different channels, colocated case

objective function in a rather complex way. This means that in some cases the solution is to be sought by using a trial-and-error approach, by varying the weights and the coefficients c_i until a satisfactory result is obtained. For instance, if only the first mode has to be damped, the choice of $c_1 = 1$ produces the results in Figure 5.21, where it is apparent that not only the first mode has been damped, but also that some modes are rather "sensitive" to the control action.

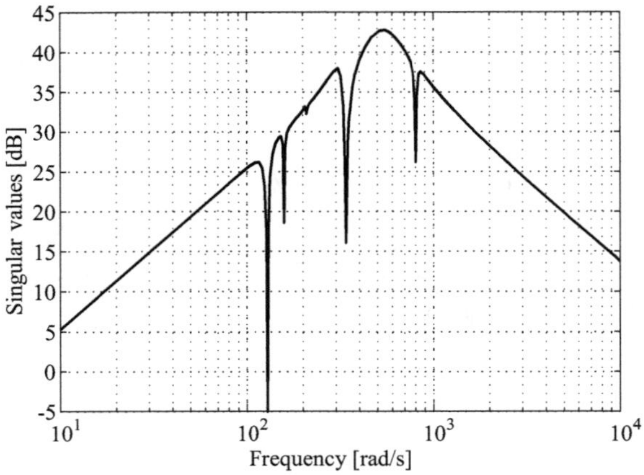

Figure 5.18 Controller principal gains, non-colocated case

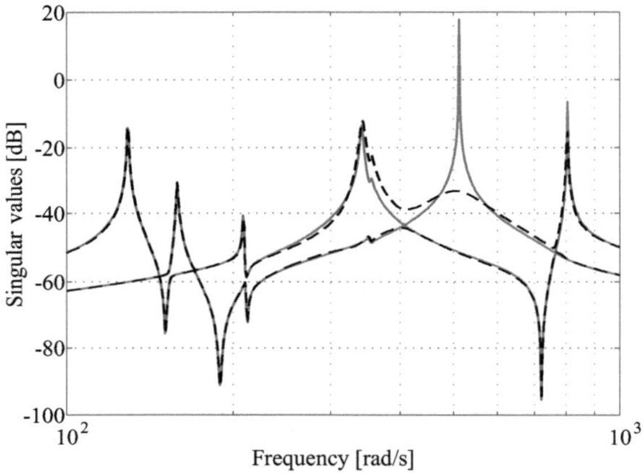

Figure 5.19 Closed loop (*dashed black lines*) and open loop (*solid grey lines*) performance, non-colocated case

A last consideration is in order in this section. The approach here proposed generalises the one in Section 5.6, which can be deduced from the former by letting $c_i = 1$ for all i, and $R_{2_1} \gg R_{2_2}$. Thus the approach can be used also in the case of high modal densities, when selecting single modes makes no sense. For this kind of systems, a sensible approach is to improve efficiency in rejecting unwanted vibration, and this can be accomplished with the considerations developed in the following section.

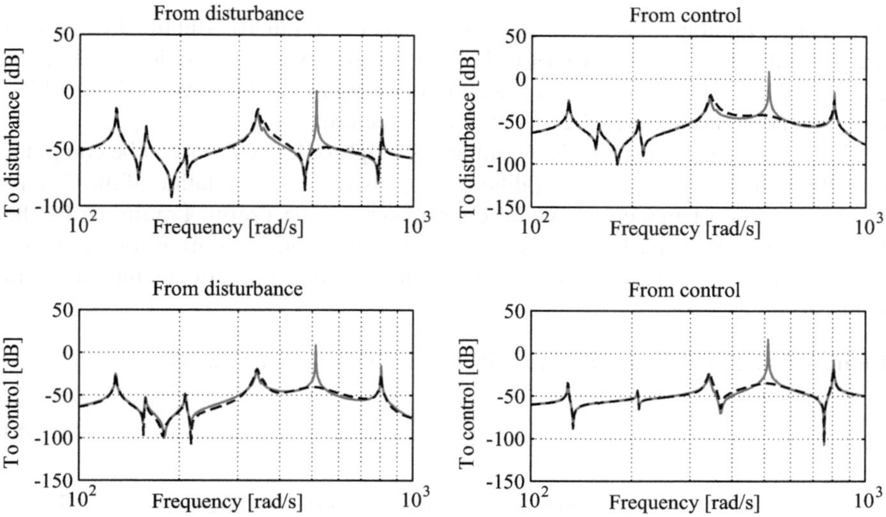

Figure 5.20 Closed loop (*dashed black lines*) and open loop (*solid grey lines*) performance for different channels, non-colocated case

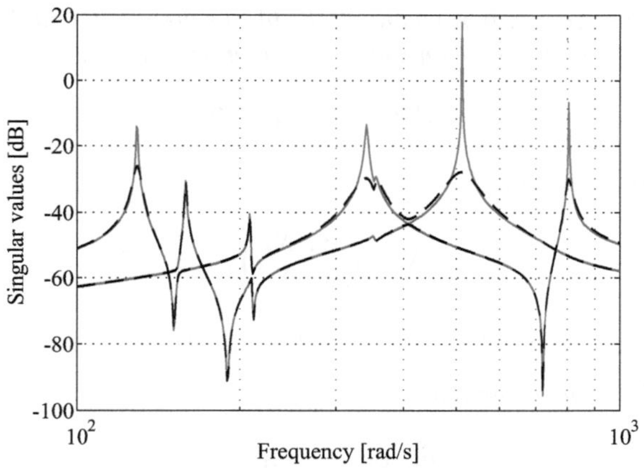

Figure 5.21 Closed loop (*dashed black lines*) and open loop (*solid grey lines*) performance

5.10.2 Computing the Gramians

In Section 5.9.1 it was shown that the optimal choice of the input scaling matrices is performed based on the controllability Gramian of the open loop and closed loop systems. However, it is clear that the efficiency of the control action depend on both the controllability and the observability Gramians. As an example, a well known strategy for model order reduction [18] consists of the computation of a balanced realisation and a reduction based on neglecting the modes associated to the Hankel singular values, *i.e.*, the singular values of a matrix obtained from both the controllability and observability Gramians, with lowest magnitude. This kind of consideration leads to the idea of selecting the weights on the matrices X_{22} and Y_{22} based on both Gramians of the model of the structure.

Since however computing the Gramians for high-order systems can be computationally burdensome, an explicit (although approximate) computation of the Gramians in the case of models in modal coordinates can be useful. Preliminarily, the following lemma is needed in order to compute the explicit solution for Lyapunov equations involving systems whose dynamic matrix can be partitioned into diagonal matrices.

Lemma 5.4. *Consider a matrix $\Delta \in \mathbb{R}^{2n\times 2n}$ with the following structure:*

$$\Delta = \begin{pmatrix} 0 & \Delta_1 \\ \Delta_2 & \Delta_3 \end{pmatrix}, \tag{5.145}$$

where $\Delta_i \in \mathbb{R}^{n\times n}, i = 1,\ldots,3$ are diagonal matrices with jth diagonal entries $\delta_{i_j}, i = 1,\ldots,3, j = 1,\ldots,n$. Let

$$Q = \begin{pmatrix} 0 & 0 \\ 0 & \hat{Q} \end{pmatrix} \tag{5.146}$$

be a $2n\times 2n$ symmetric matrix, where all the blocks are $n\times n$, and denote by $\hat{q}_{ij}$ the (i,j)th entry of the matrix $\hat{Q}$. Consider the Lyapunov equation in the unknown X

$$\Delta^{\mathrm{T}}X + X\Delta + Q = 0, \tag{5.147}$$

partition X into $n\times n$ matrices

$$X = \begin{pmatrix} X_1 & X_2 \\ X_2^{\mathrm{T}} & X_3 \end{pmatrix}, \tag{5.148}$$

and denote the (i,j) entry of $X_k, k = 1,\ldots,3$ by $x_{k_{ij}}$. Then

1. *Equation 5.147 has a unique solution iff*

$$\delta_{3_{i,j}}\delta_{123_{i,j}} \neq \delta^2_{12_{i,j}}, \forall i,j, \tag{5.149}$$

where

$$\begin{aligned}
\delta_{3_{i,j}} &= \delta_{3_i} + \delta_{3_j}, & (5.150)\\
\delta_{12_{i,j}} &= \delta_{1_i}\delta_{2_i} - \delta_{1_j}\delta_{2_j}, & (5.151)\\
\delta_{123_{i,j}} &= \delta_{1_i}\delta_{2_i}\delta_{3_j} + \delta_{1_j}\delta_{2_j}\delta_{3_i}. & (5.152)
\end{aligned}$$

2. *If the above condition holds, the (i,j)th entries of the matrices X_k depend only on the elements δ_{k_i}, δ_{k_j}, $k = 1,\ldots,3$, $i,j = 1,\ldots,n$.*
3. *The above entries are explicitly given by*

$$x_{1_{ij}} = \frac{\delta_{2_i}\delta_{2_j}}{\delta_{123_{i,j}} - \frac{\delta^2_{12_{i,j}}}{\delta_{3_{i,j}}}}\hat{q}_{ij}, \tag{5.153}$$

$$x_{2_{ij}} = \frac{\delta_{2_i}}{\delta_{123_{i,j}}\frac{\delta_{3_{i,j}}}{\delta_{12_{i,j}}} + \delta_{12_{i,j}}}\hat{q}_{ij}, \tag{5.154}$$

$$x_{3_{ij}} = \frac{1}{\frac{\delta^2_{12_{i,j}}}{\delta_{123_{i,j}}} - \delta_{3_{i,j}}}\hat{q}_{ij}. \tag{5.155}$$

Proof. Equation 5.147 can be written as

$$\Delta_2 X_2^{\mathrm{T}} + X_2\Delta_2 = 0, \tag{5.156}$$
$$\Delta_2 X_3 + X_1\Delta_1 + X_2\Delta_3 = 0, \tag{5.157}$$
$$X_3\Delta_2 + \Delta_1 X_1 + \Delta_3 X_2^{\mathrm{T}} = 0, \tag{5.158}$$
$$\Delta_1 X_2 + X_2^{\mathrm{T}}\Delta_1 + \Delta_3 X_3 + X_3\Delta_3 = -\hat{Q}, \tag{5.159}$$

by defining the new variable $\bar{X}_2 = X_2\Delta_2$, which, according to Equation 5.156 is an antisymmetric matrix, and with simple manipulations, the above system can be simplified as

$$X_1\Delta_1\Delta_2 + \bar{X}_2\Delta_3 + \Delta_2 X_3\Delta_2 = 0, \tag{5.160}$$
$$\Delta_1\Delta_2 X_1 - \Delta_3\bar{X}_2 + \Delta_2 X_3\Delta_2 = 0, \tag{5.161}$$
$$\Delta_1\Delta_2\bar{X}_2 - \bar{X}_2\Delta_1\Delta_2 + \Delta_2\Delta_3 X_3\Delta_2 + \Delta_2 X_3\Delta_2\Delta_3 = -\Delta_2\hat{Q}\Delta_2. \tag{5.162}$$

By defining the unit basis vector e_i as the n-vector whose only non-null entry is 1 at ith entry and premultiplying by e_i^{T} and post-multiplying by e_j Equations 5.160–5.162 results in

$$\delta_{1_j}\delta_{2_j}x_{1_{ij}} + \delta_{3_j}\bar{x}_{2_{ij}} + \delta_{2_i}\delta_{2_j}x_{3_{ij}} = 0, \tag{5.163}$$
$$\delta_{1_i}\delta_{2_i}x_{1_{ij}} - \delta_{3_i}\bar{x}_{2_{ij}} + \delta_{2_i}\delta_{2_j}x_{3_{ij}} = 0, \tag{5.164}$$
$$\delta_{12_{i,j}}\bar{x}_{2_{ij}} + \delta_{2_i}\delta_{2_j}\delta_{3_{i,j}}x_{3_{ij}} = -\delta_{2_i}\delta_{2_j}\hat{q}_{ij}. \tag{5.165}$$

Replacing $\bar{x}_{2_{ij}} = \delta_{2_j}x_{2_{ij}}$ and solving the above equations for all i, j proves the lemma. □

As a consequence of the above lemma, two corollaries are deduced for an explicit solution of the controllability and observability Gramians.

Corollary 5.3. *For a flexible system in the modal form 5.3–5.7 the controllability Gramian is given by*

$$P = \begin{pmatrix} P_1 & P_2 \\ P_2^{\mathrm{T}} & P_3 \end{pmatrix}, \tag{5.166}$$

where P_2 is antisymmetric, and the elements of the submatrices are given by

$$p_{1_{ij}} = \frac{\lambda_i + \lambda_j}{d_{ij}} \hat{b}_{X_{ij}}, \tag{5.167}$$

$$p_{2_{ij}} = \frac{\omega_{n_i} - \omega_{n_j}}{d_{ij}} \hat{b}_{X_{ij}}, \tag{5.168}$$

$$p_{3_{ij}} = \frac{\lambda_i \omega_{n_j} + \lambda_j \omega_{n_i}}{d_{ij}} \hat{b}_{X_{ij}}, \tag{5.169}$$

where $\hat{b}_{X_{ij}}$ is the (i,j)th element of the matrix $\hat{B}_X$ in (5.131) and

$$d_{ij} = (\omega_{n_i} - \omega_{n_j})^2 + (\lambda_i + \lambda_j)(\lambda_i \omega_{n_j} + \lambda_j \omega_{n_i}). \tag{5.170}$$

Proof. It is sufficient to apply Lemma 5.4 to system 5.3–5.7. □

Corollary 5.4. *For a flexible system in the modal form 5.3–5.7 the observability Gramian is given by*

$$Q = \begin{pmatrix} -\Omega & 0 \\ 0 & I_{n\times n} \end{pmatrix} P_c \begin{pmatrix} -\Omega & 0 \\ 0 & I_{n\times n} \end{pmatrix}, \tag{5.171}$$

where P_c is the controllability Gramian of the pair (A,C).

Proof. Applying Lemma 5.4 to system 5.3–5.7 and partitioning Q in $n \times n$ submatrices

$$Q = \begin{pmatrix} Q_{11} & Q_{12} \\ Q_{21} & Q_{22} \end{pmatrix}, \tag{5.172}$$

the elements of such submatrices are given by

$$q_{1_{ij}} = \frac{\omega_{n_i} \omega_{n_j} (\lambda_i + \lambda_j)}{d_{ij}} \hat{b}_{Y_{ij}}, \tag{5.173}$$

$$q_{2_{ij}} = \frac{\omega_{n_i} (\omega_{n_j} - \omega_{n_i})}{d_{ij}} \hat{b}_{Y_{ij}}, \tag{5.174}$$

$$q_{3_{ij}} = \frac{\lambda_i \omega_{n_j} + \lambda_j \omega_{n_i}}{d_{ij}} \hat{b}_{Y_{ij}}, \tag{5.175}$$

where $\hat{b}_{Y_{ij}}$ is the (i,j)th entry of the matrix $\hat{B}_Y$ in Equation(5.132. By using Corollary 5.3 the theorem is easily proved. □

It is interesting to point out that for systems with low modal density (*i.e.*, with few modes in a narrow frequency intervals) and low damping, so that the modes are well "separated" in frequency, the terms $(\omega_{n_i} - \omega_{n_j})^2$ dominate in Equations 5.167–5.169. This motivates the definitions of system with well separated modes given at the beginning of the previous chapter.

Systems with well separated modes have interesting approximation properties, as shown by the following theorem.

Theorem 5.6. *For systems with well separated modes the control Gramian is diagonal dominant and can be effectively approximated with*

$$P \approx \begin{pmatrix} \mathrm{diag}(\Lambda^{-1}B_X\Omega^{-1}) & 0 \\ 0 & \mathrm{diag}(\Lambda^{-1}B_X) \end{pmatrix}. \tag{5.176}$$

Proof. From expressions 5.167 and 5.169 it is clear that the elements of the block matrices P_1 and P_3 defining the Gramian are the elements of the matrix $\hat{B}_X$ multiplied by "weighting" functions assuming their maximum value on the leading diagonal. Based on physical considerations, it is reasonable to assume that the entries of $\hat{B}_X$ do not differ very much from each other, since the entries of $\hat{B}_X$ in modal variables are obtained from the normalised mode shapes at the abscissa where the actuator acts, as shown in Chapter 2, hence the matrices P_1 and P_3 can be approximated by their diagonal elements. Moreover, since P_2 is antisymmetric, its leading diagonal is zero. □

Note that the same approximation is not acceptable for the observability Gramian, as the elements of Q_2 cannot be neglected, as apparent from Equation 5.174.

However, the product PQ, whose eigenvalues are computed in order to find the Hankel singular values, can be approximated by a diagonal matrix. This is the result of the following theorem.

Theorem 5.7. *The Hankel singular values of a flexible system with well separated modes can be approximated by the following expression:*

$$\sigma_i = \frac{1}{2\lambda_i}\sqrt{\bar{b}_{X_{ii}}\bar{b}_{Y_{ii}}},\ i = 1,\ldots,n, \tag{5.177}$$

where $\bar{b}_{X_{ii}}$ and $\bar{b}_{Y_{ii}}$ are the diagonal entries of the matrices

$$\bar{B}_X = \{\hat{b}_{X_{ij}}/d_{ij}\}, \tag{5.178}$$

$$\bar{B}_Y = \{\hat{b}_{Y_{ij}}/d_{ij}\}. \tag{5.179}$$

Proof. Computing the product $M = PQ$, and partitioning M into $n \times n$ matrices

$$M = \begin{pmatrix} M_1 & M_2 \\ M_3 & M_4 \end{pmatrix}, \tag{5.180}$$

simple algebraic computations show that the elements of the matrix M_2 can be written as

$$m_{2_{ij}} = \sum_{k=1}^{n} \frac{(\lambda_i+\lambda_k)(\omega_{n_j}-\omega_{n_k})\omega_{n_k} + (\omega_{n_i}-\omega_{n_k})(\lambda_k\omega_{n_j}+\lambda_j\omega_{n_k})}{d_{ik}d_{kj}}\bar{b}_{X_{ik}}\bar{b}_{Y_{kj}}, \tag{5.181}$$

which, in the hypothesis of well separated modes, can be neglected. Hence the matrix M is lower triangular. Moreover, the same approximation shows that both M_1 and M_4 are diagonally dominant. Thus the result of the theorem is simply an approximation of the product $p_{1_{ii}}q_{1_{ii}}$. □

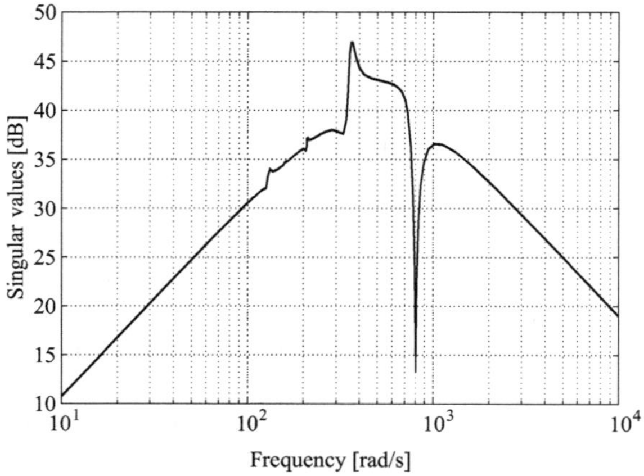

Figure 5.22 Controller principal gains

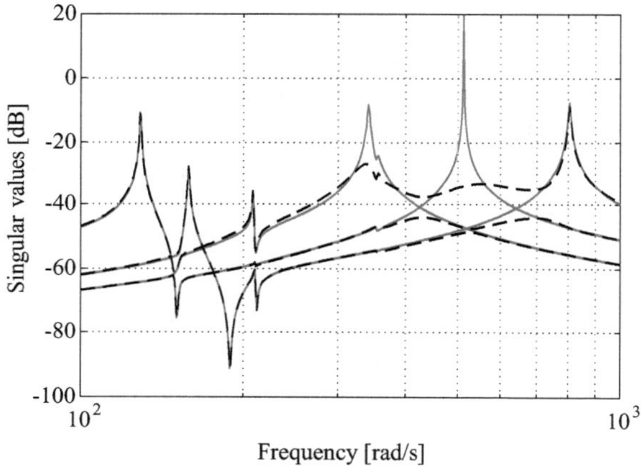

Figure 5.23 Closed loop (*dashed black lines*) and open loop (*solid grey lines*) performance

From the proof of the theorem it is clear that the order of approximation can be assessed by checking the dominance of the terms $(\omega_{n_i} - \omega_{n_j})^2, \forall i, j$.

Based on the above approximation, it is reasonable to choose the weights c_i proportionally to the Hankel singular values just computed. This approach is tested in simulation in the following. Again the same structure is considered with the very unfavourable case of two disturbance and one control input only. First, the case with scaling matrices as in Equation 5.144 is considered. The results are shown in Figures 5.22–5.24.

Also in this case, two situations are considered: colocated and non-colocated feedback. For the colocated feedback, the results are shown in Figures 5.25 and 5.26. The singular values of the controller are not shown for brevity. For the non-colocated

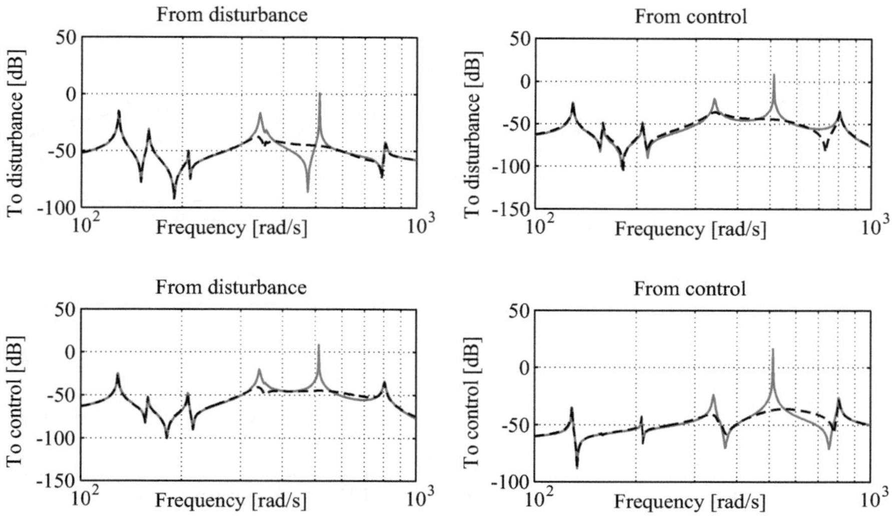

Figure 5.24 Closed loop (*dashed black lines*) and open loop (*solid grey lines*) performance from different channels

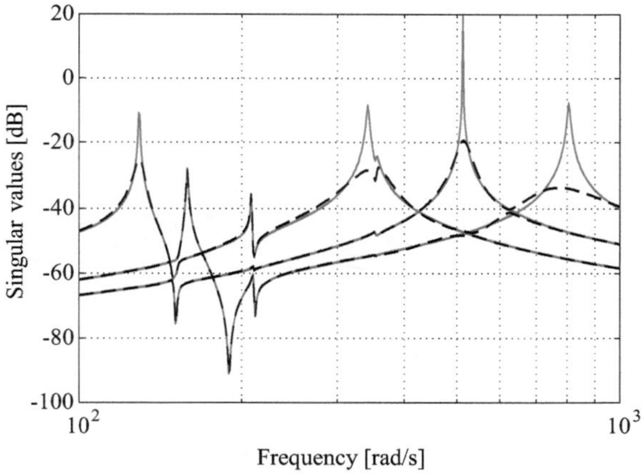

Figure 5.25 Closed loop (*dashed black lines*) and open loop (*solid grey lines*) performance

feedback, the results are shown in Figures 5.27 and 5.28. The singular values of the controller are not shown for brevity.

The simulations show that, although the colocated case works better, the non-colocated case is better able to counteract the effect of disturbances entering the structure at the locations where the non-colocated sensors are placed. Thus, the following strategy can be deduced: to reject disturbances at a particular location simply add a sensor there. Experimental results will be presented and discussed in Chapter 7.

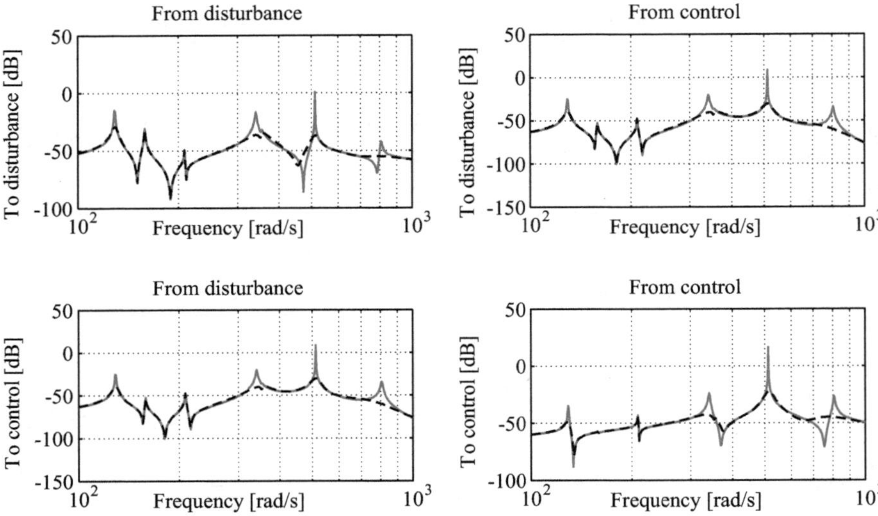

Figure 5.26 Closed loop (*dashed black lines*) and open loop (*solid grey lines*) performance from different channels

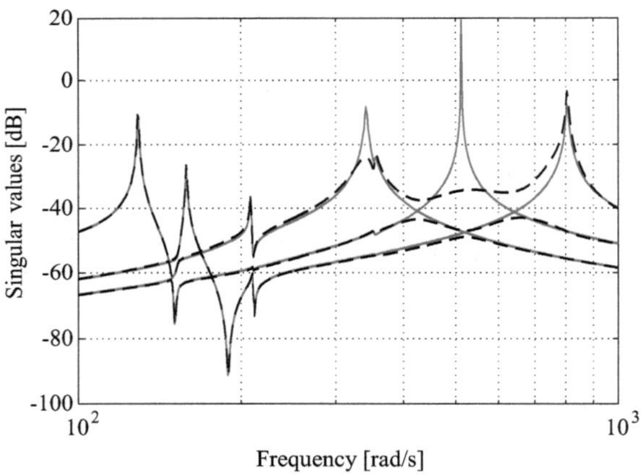

Figure 5.27 Closed loop (*dashed black lines*) and open loop (*solid grey lines*) performance

5.11 Conclusions

This chapter addressed the active vibration control in the most complex case, *i.e.*, a flexible structure with high modal density subject to broadband disturbances. The approach followed in the chapter is the use of optimal control strategies, exploiting specific solutions of the general optimal control problem based on the properties of the dynamic model derived in Chapter 2. The first proposed solution resulted in a closed-form solution to an $\mathscr{H}_\infty$ control problem. Its well known robustness

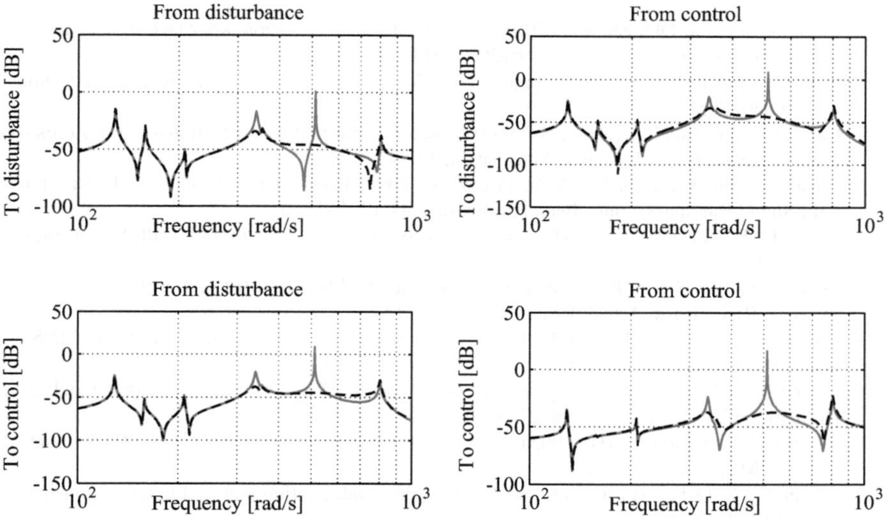

Figure 5.28 Closed loop (*dashed black lines*) and open loop (*solid grey lines*) performance from different channels

properties make the strategy very attractive for high-order systems such as flexible structures with high modal density subject to broadband disturbances. The same approach can obviously be extended to the $\mathscr{H}_2$ case, as it is well known that a $\mathscr{H}_2$ is a particular solution of an $\mathscr{H}_\infty$ problem when the limit on the infinity-norm is left unbounded. However, an $\mathscr{H}_2$ strategy allows generalisations in the procedure that are not feasible for the general $\mathscr{H}_\infty$ case. In particular, an optimal selection of the scaling matrix selection in the $\mathscr{H}_2$ case is possible. Moreover, the $\mathscr{H}_2$ control strategy can be extended to the selective reduction of specific modes of the structure. Finally, the use of non-square controllers was explored, based on the physical consideration that employing more sensors (accelerometers) is easier and cheaper than adding actuators. Solving this problem required the solution of an LMI problem. All proposed controllers have been tested with numerical examples and will be experimentally implemented in Chapter 7.

References

1. Arheli R, Johnson MA and Wilkie J (1997) LQG Stable Stabilizing Control: Some Recent Results. In: Proc of the 36th IEEE Conference on Decision and Control, San Diego, CA
2. Balas GJ, Doyle JC (1994) Robustness and Performance Trade-offs in Control Design for Flexible Stuctures. IEEE Trans Contr Syst Techn, 2:352–361
3. Blondel V, Lundvall C (1995) A Rational Test for Strong Stabilization. Automatica, 31:1197–1198

4. Campos-Delgado DU, Zhou K (2003) A Parametric Optimization Approach to $\mathscr{H}_\infty$ and $\mathscr{H}_2$ strong stabilization. Automatica, 39:1205–1211
5. Cao YY, Lam J (2000) On Simultaneous H_∞ Control and Strong H_∞ Stabilization. Automatica, 36:859–865
6. Calafiore G, Carabelli S, Bona B (1997) Structural Interpretation of Transmission Zeros for Matrix Second-Order Systems. Automatica, 33:745–748
7. Cavallo A, De Maria G, Natale C, Pirozzi S, (2008) Robust Control of Flexible Structures with Stable Bandpass Controllers. Automatica, 44:1251–1560.
8. Dorf RC, Bishop RH (2001) Modern Control Systems. 9th Ed, Prentice Hall, Upper Saddle River, NJ.
9. Dunn W (1994) Signal Processing for Micromachined Sensors. In: Sensor Technology and Devices, Ristic L Ed, Artech House Norwood, MA
10. Emami-Naeini A, Van Dooren P (1982) Computation of Zeros of Linear Multivariable Systems. Automatica, 18:415–430
11. Falangas ET, Dworak JA, Koshigoe S (1994) Controlling Plate Vibrations Using Piezoelectric Actuators. IEEE Control Systems, 8:34–41
12. Fuller CR, von Flotow AH (1995) Active Control of Sound and Vibrations. IEEE Control Systems Magazine. 15:9–19
13. Gümüssoy S, Özbay H (2005) Remarks on Strong Stabilization and Stable $\mathscr{H}_\infty$ Control Design. IEEE Trans Autom Contr, 50:2083–2087
14. Halevi Y (1994) Stable LQG Controllers. IEEE Trans Autom Contr, 39:2104–2106
15. Junkins JL, Kim Y (1993) Introduction to Dynamics and Control of Flexible Structures. AIAA Education Series, Washington DC
16. Kar IN, Miyakura T, Seto K (2000) Bending and Torsional Vibration Control of a Flexible Plate Structure using $\mathscr{H}_\infty$-Based Robust Control Law. IEEE Trans Contr Syst Tech, 8:545–553
17. Maciejowski J.M. (1989) Multivariable Feedback Design. Addison-Wesley, Wokingham
18. Moore BC (1981) Principal Components Analysis in Linear Systems: Controllability, Observability and Model Reduction. IEEE Trans Autom Contr, 26:17–32
19. Olgac N, Hosek M (1997) Active Vibration Absorption Using Delayed Resonator with Relative Position Measurement. ASME J Vibr Acoust, 119:131-136
20. Youla DC, Bongiorno JJ, Lu CN (1974) Single-loop Stabilization of Linear Multivariable Plants. Automatica, 10:159–173
21. Zeren M, Özbay H (1997) On Stable $\mathscr{H}_\infty$ Controller Design. In: Proc of the IEEE Amer Contr Conf, Albuquerque, NM
22. Zeren M, Özbay H. (1999) On the Synthesis of Stable $\mathscr{H}_\infty$ Controllers. IEEE Trans Autom Contr, 44:431–435
23. Zeren M, Özbay H (2000) On the Strong Stabilization and Stable $\mathscr{H}_\infty$ Controller Design Problems for MIMO Systems. Automatica, 36:1675–1684
24. Zhou K, Doyle JC (1998) Essentials of Robust Control. Prentice Hall, Upper Saddle River, NJ

Chapter 6
Implementation Issues

6.1 Introduction

Actuators play a critical role in active structures technology and, in many cases, they constitute the limiting factor. In the near future, the conventional actuators will still be the most important. However, in some cases, they seem to have difficulty satisfying new and advanced demands. In such cases, smart material-based devices may offer a satisfactory solution. Nevertheless, as is widely known, smart actuators on the one hand provide a high potential while, on the other hand, they may complicate the control design because of their typical nonlinear characteristics. As a consequence, the assumption of linearity, underlying identification and control techniques presented in previous chapters, is often not verified when smart actuators and/or sensors are used in the control system. When the devices' nonlinearities are not negligible, the use of smart systems to implement linear active control strategies needs an accurate modelling and a suitable control. For these reasons, the selection of proper actuators is more complicated than the selection of the sensors, primarily due to their effect on the dynamic behaviour of the overall system. Furthermore, the selection of the actuators affects both the power needs and the coupling mechanisms between the structure and the actuator. The coupling mechanism must usually be designed according to the particular actuator technology and the constraints imposed by the structure. Finally, the choice of an actuator for a specific task must be made by also taking into account some typical performance parameters, such as the continuous power output, the range of motion, the resolution, the accuracy, the frequency response, the power requirement, the peak force/torque, the heat dissipation and the speed characteristics. This chapter provides technical information useful for selecting the type of actuator and/or sensor for the implementation of a vibration control system and some techniques that allow optimising performance by properly using non-conventional devices. The described actuators together with the rapid control prototype system, presented in the latter section of this chapter, are used in the experiments illustrated in Chapter 7.

6.2 Conventional Actuators

A *hydraulic actuator* generally consists of a piston in a hydraulic cylinder that has hydraulic fluid openings at each end. When high pressure fluid is introduced into one end of the cylinder, the piston moves to the other end, expelling fluid from the opening. When the supply of high pressure fluid is switched to the other end of the cylinder the piston moves to the opposite direction. The most common method to alternate the hydraulic supply from one end to the other is to use a servo-valve consisting of a moveable spool connected to a solenoid. Servo-valves can be quite expensive, depending on the upper frequency at which they are required to operate, which can be 3 Hz to 20 Hz for an active vehicle suspension or 150 Hz for structural vibration control. One advantage of hydraulic actuators is their large displacement and high force generating capability with a relatively small size. Disadvantages include the need for a hydraulic power supply, which can be inherently noisy and non-linearities between the servo-valve input voltage and the hydraulic actuator force or displacement output.

Pneumatic actuators are very similar in operation to hydraulic actuators, except that the hydraulic fluid is replaced by air. One advantage of pneumatic actuators is that they can use the same air supply as passive air springs, which may mounted in parallel with them. The major disadvantage of pneumatic actuators is their relatively low bandwidth (less than 10 Hz) due to the compressibility of the air.

The *proof mass actuator* consists of a mass that is free to slide along a track. The mass can be accelerated using an electromagnetic field or by connecting it to a ball screw. In these cases the proof mass results in a reaction in the ball screw or electromagnetic stator, which are attached to the structure to be controlled. One problem with linear proof mass actuators is the limited motion of the reaction mass, which becomes more serious at low frequencies. However, the problem is much less severe for the ball screw type actuator as high forces can be produced on the reaction mass without producing large motions.

Electrodynamic actuators or shakers consist of a moveable core to which a cylindrical electrical coil is fixed. The core and coil move back and forth in magnetic field that is generated by a permanent magnet (smaller shakers) or by a DC current flowing through a second coil fixed to the stationary portion of the shaker (stator). The moveable core is supported by frames, which are very flexible in the axial direction in which it is desired that the core move and rigid in the radial direction to prevent the core touching the outer armature. To maximise lightness and stiffness, the core is usually made of high strength aluminium alloy and is usually hollow. When a sinusoidal voltage is applied to the drive coil, the polarity and strength of the drive coil magnetic field changes in phase with the applied voltage and so then does the force of attraction between the field coil (or permanent magnet) and the driving coil, resulting in axial movement of the core (or armature). The commercially available shaker can generate sinusoidal forces from few Newtons to hundreds of kilonewtons in a typical frequency range of $[2, 10000]$ Hz. The smallest models weigh no less than 1 kg and are quite cumbersome. Electromagnetic actuators are similar in con-

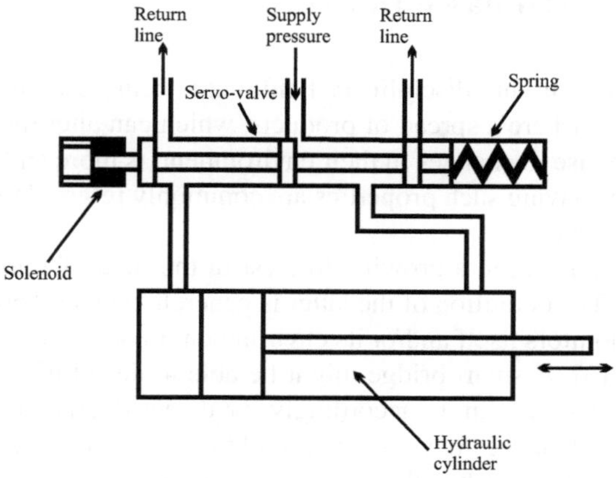

(a)

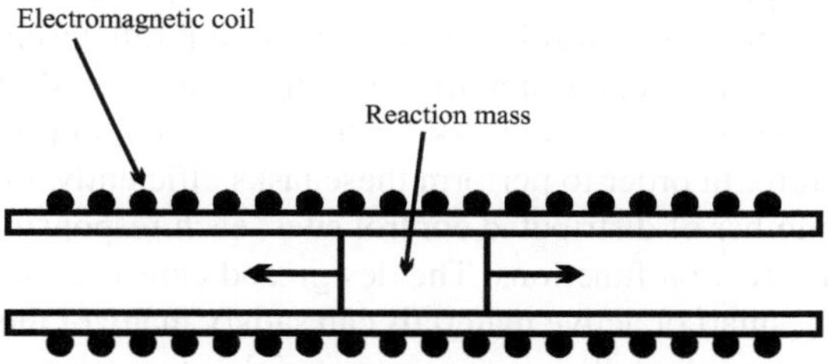

(b)

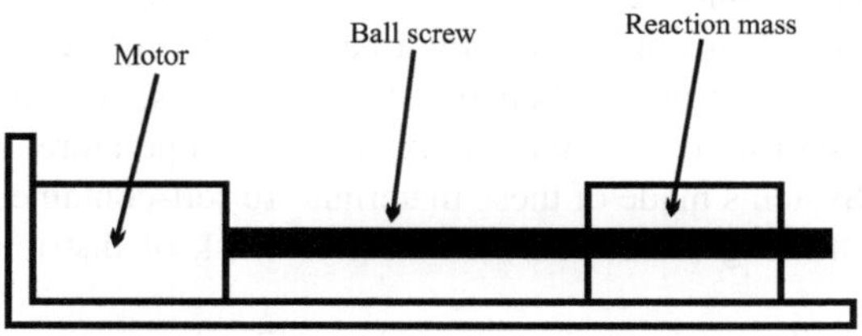

(c)

Figure 6.1 (a) Schematic of a hydraulic actuator driven by a servo-valve, (b) of a proof mass actuator with electromagnetic drives and (c) of a proof mass actuator with a ball screw drive

struction to electrodynamic shakers except that the inner core as well as the armature is fixed. The Figure 6.1 reports some schemes of typical conventional actuators.

6.3 Smart Material-based Devices

Smart technology is an interdisciplinary field that is being used in a wide range of application sectors where a spread of products, which can alter their physical characteristics in response to changes in their environment, is more and more frequently proposed. Devices having such properties are commonly referred to as "smart", "intelligent" or "adaptive".

The last years have seen a growing interest in the field of *smart materials* and *smart structures*. The definition of the latter is generally agreed upon: a smart structure is one that monitors itself and/or its environment in order to respond to changes in its condition [13]. A smart bridge might be able to detect high wind conditions and reduce posted speed limits accordingly, or detect degradation in footings or piers and alert maintenance crews. A smart building may sense the presence of high winds or the onset of an earthquake and adjust its dynamic properties automatically to minimise damage or occupant discomfort. To a smart aeronautic or aerospace systems, control transducers and actuators should be lightweight, should use the available power supplies and in some cases should be capable of multiple roles. For example, the actuators that control vibrations in an aircraft fuselage may also act as inputs and sensors for health monitoring of the structure. Moreover, in certain applications, the same devices may work for the scavenging of power from components of the structure. In order to perform these tasks efficiently, the smart structure requires a large number of distributed control sites, each responsible for local sensing, decision and actuation functions. The design and construction of actuators and sensors employing smart or active materials can satisfy many of these requirements.

Just what constitutes "intelligence" in a material is the subject of much debate. Some researchers argue that there is no single material that satisfies the dictionary definition of "intelligent", although hybrid materials can come close [13]. However, there are a number of materials that might be termed "self-adapting" or "responsive", and the ability to adapt is certainly a part of "intelligence". These materials are able to alter one or more of their physical properties, such as stiffness, strain or viscosity, in response to external stimuli such as temperature, stress, electric or magnetic fields. Systems made of these materials are self-contained and often completely scalable, making them well-suited for the task of distributed sensing and actuation.

The implementation of control systems that utilise innovative devices based on these materials requires the development of models that quantify their interaction with controlled structures. In particular, all active materials exhibit nonlinear dynamics and hysteresis at high drive levels. As a consequence the devices, exploiting the advantages of these materials, generally suffer from nonlinear characteristics. The nonlinearities must be taken into account in both models and control methods before the controlled systems can achieve full performance. The control design for structures that use nonlinear and hysteretic actuators or sensors can involve different approaches ranging from linear robust control employing approximate linear models to fully nonlinear control laws that directly accommodate the nonlinearities inherent to the devices. The compensation of undesired nonlinear characteristics improves the performance of applications utilising smart material based devices.

A lot of smart materials have been developed in recent years and many of them are described in [30, 31]. The most interesting materials for applications to flexible structures are piezoelectric ceramics and magnetostrictive materials. The first allow us to design actuators that are lightweight, space-saving and inexpensive for different applications. The latter enable us to generate high forces associated with high precision deformations in both static and dynamic work conditions. The magnetostrictive actuators that fully exploit the material potential generally show a greater hysteresis than piezoelectric ones, but they offer the same performances in terms of generated forces with much lower voltages. Furthermore, devices based on these materials can work as actuators and/or sensors at the same time, offering the possibility to be used in advanced control techniques as self-sensing devices and to operate in different roles depending on the needs.

6.3.1 Piezoelectric Ceramics

Piezoelectricity was discovered by Curie brothers in 1880, who initially worked on tourmaline and later on quartz and other materials. They were able to demonstrate the generation of electric potential in response to applied pressure or stresses. This is the direct piezoelectric effect, while the converse effect, *i.e.*, the generation of strains or displacement in the material in response to applied fields, was discovered later from thermodynamic principles. Both effects depend on the non-centrosymmetric nature of certain ceramics. Since then several piezoelectric materials have been developed, initially based on single crystal ceramics and later on polycrystalline ceramics. The latter have several advantages in terms of higher sensitivity and ease of fabrication into a variety of shapes and sizes. Nowadays, in different application fields, the most commonly used piezoelectric ceramics are barium titanate ($BaTiO_3$), lead zirconate titanate (PZT) and lead magnesium niobate. Piezoelectric materials are widely considered for innovative actuators and sensors design due to the fact that the resulting devices are lightweight, not very expensive, compact and have broadband drive capabilities and very high set point accuracy. Furthermore, the direct and converse piezoelectric effects allow these devices to work simultaneously as sensors and actuators. The piezoelectric materials exhibit hysteresis and constitutive nonlinearities at all drive levels owing to their non-centrosymmetric nature. These effects can be mitigated through feedback mechanisms or simply neglected for low drive regimes. Instead, in high drive levels, it is necessary to have an accurate model of the device and to design a suitable control system that takes into account hysteresis and other nonlinearities, in order to achieve the maximum performances. The commercial piezoelectric actuators can be either monolithic or made of thin stacked layers. Each layer can have a thickness of 30–200 μm. Both types of actuators can generally exhibit a strain up to 0.2% with a supply voltage higher in the monolithic case, but always of the order of hundreds volts. Figure 6.2 shows a typical quasi-

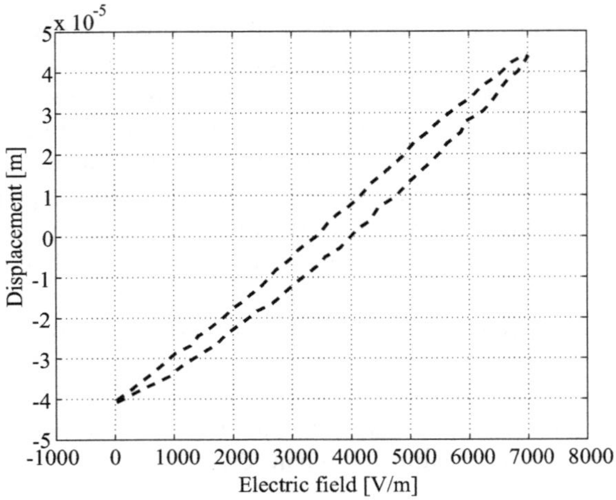

Figure 6.2 A typical electric field/diplacement relation of a piezoelectric actuator

static field-displacement relation for a PZT stacked actuator. Piezoelectric actuators are currently used in different application fields including precision positioning, vibration suppression, noise control, dot matrix printer heads, alarm systems, medical sensors, industrial sensors and many other commercial and prototypical applications.

6.3.2 Magnetostrictive Materials

In 1842 James P. Joule began the research on magnetostrictive materials measuring a change in the length of an iron sample subjected to a magnetic field. This change in dimension due to a change in magnetisation is phenomenologically analogous to the converse piezoelectric effect and it is known as the Joule effect. Soon after the Villari or magnetostrictive effect was discovered, that is the reciprocal behaviour in which stresses produce a change in magnetisation. The latter is phenomenologically similar to the direct piezoelectric effect. As a consequence, magnetostrictive materials can be used for the design of actuators and/or sensors. The first applications were very limited since the materials exhibited a maximum strain of the order of 50 $\mu\varepsilon$. In the second half of the 20th century "giant" magnetostrictive materials were discovered and investigated. In particular, in 1963 magnetostrictive alloys comprised of rare earth elements (like terbium and dysprosium) were able for the first time to produce strains on the order of 1000 $\mu\varepsilon$ at cryogenic temperature. Then almost simultaneously some researchers of the Naval Ordnance Laboratory (NOL) and the Naval Research Laboratory (NRL) demonstrated that the combination of these rare earth alloys with the transition metal iron produce alloys that show giant magnetostrictive properties at room temperatures. Thus the Terfenol-D (ter: terbium, fe: iron, nol:

NOL, D: dysprosium) was developed, which exhibits strain up to 1800 $\mu\varepsilon$ at room temperature and can achieve 4000 $\mu\varepsilon$ at resonance. The analogy with piezoelectric materials gives the same interest to the magnetostrictive materials for designing innovative actuators and/or sensors in advanced applications. Magnetostrictive actuators proved to be very attractive due to their large strain, high force and high energy density. It was immediately realised that their use for controlling noise and vibration in flexible structures was a quite promising application field. Due to the circuits required to generate the driving magnetic fields, devices based on magnetostrictive materials are currently larger and more cumbersome than piezoelectric patches. However, magnetostrictive transducers also have very interesting advantages for certain applications, such as giant forces and low voltage supply (of the order of tens of volts). Hence, if the material properties and the design of the device are refined, their application fields may rapidly increase. From the modelling and control point of view, magnetostrictive-based devices present not negligible hysteresis and nonlinearities even at moderate drive levels. Furthermore, the hysteresis of a magnetostrictive actuator is quite different than the classical one since its characteristic presents only positive strain. Figure 6.3 reports the typical butterfly-shaped hysteresis cycle of a magnetostrictive actuator. Clearly, it is possible to obtain the full displacement range by limiting the current to positive values. As a consequence, it is necessary to take into account these unwanted nonlinear characteristics in the control system design in order to use the material at full potential. Nowadays, other magnetostrictive materials have been developed by including holmium in the produced alloy or replacing the terbium with the gallium. The new alloys are aimed at reducing hysteresis and material anisotropy, but Terfenol-D is still the most commonly used magnetostrictive material. Applications that utilise Terfenol-D alloys include active vibration and noise control, micropositioning in high-force regimes, medical and industrial ultrasonics applications, non-contact torque sensors, tuned and dynamic vibration absorbers. For a satisfactory review on the most recent contributions in Terfenol-D magnetostriction see [15].

6.4 Effects of Hysteresis

In previous chapters identification and control techniques were presented that model the flexible structures under the linearity assumption. This hypothesis, certainly met by flexible structures, is not verified when smart actuators and/or sensors are used in the control system. As discussed above, devices based on smart materials present nonlinear characteristics depending on drive levels. When these nonlinearities are not negligible, the use of smart transducers to implement the active control strategies presented in this book needs an accurate modelling and a suitable control. The first nonlinear effect to be discussed is hysteresis. In particular, this section proposes a control strategy to reduce the effects of hysteresis in systems employing devices based on smart materials. The procedure presented has been experimentally tested on a commercial magnetostrictive actuator.

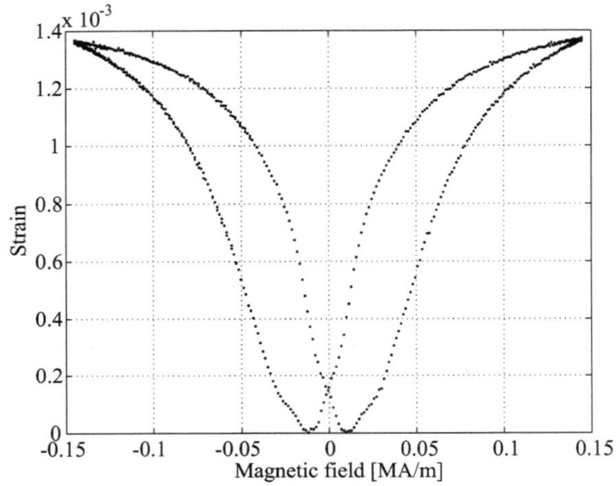

Figure 6.3 A typical magnetic field/strain relation of a magnetostrictive actuator

The problem of modelling hysteresis phenomena traditionally concerns magnetisation processes in ferromagnetic materials. Such modelling can be carried out either on a physical ground, describing processes on a mesoscopic scale [5] or, from the phenomenological viewpoint, by defining mathematical operators with memory able to describe input–output relationships of systems with hysteresis [23]. The latter is the approach proposed by the authors.

In feedback control systems, hysteresis may be an undesired effect that decreases the performance of the system and may complicate its design [10]. As a consequence, the control design may require a suitable hysteresis compensation procedure and, therefore, a satisfactory hysteresis modelling for the device. Generally speaking, an hysteresis compensator is a device implementing an inverse operator of the hysteresis. However, even if the availability of the inverse hysteresis operator is guaranteed from the mathematical viewpoint, such a compensator can be exactly formalised in very few cases.

6.4.1 Hysteresis Modelling

The well known classical Preisach model (PM) can be seen as the superposition of a continuous set of two-position relays and formalised, according to [23], as

$$y(t) \equiv \Gamma[x](t) = \iint_{\alpha \geq \beta} \mu(\alpha,\beta)\,\widehat{\gamma}_{\alpha\beta}\,x(t)\,\mathrm{d}\alpha \mathrm{d}\beta, \tag{6.1}$$

where $\widehat{\gamma}_{\alpha\beta}$ represents the relay operator (which takes the values +1 and -1), α and β being the "up" and "down" relay's switching values and $\mu(\alpha,\beta)$ is the Preisach distribution function (PDF). As it is well known, the PM fulfills some general properties, such as the *wiping-out* property, *i.e.*, given an input sequence, only the dominant extrema are stored by the PM, all other input extrema are wiped out and the *congruency* property, *i.e.*, all minor loops described by input variations within the same lower and upper extrema are congruent by vertical translation (more details can be found in [23]).

From a practical point of view, the reconstruction of the distribution function is not strictly necessary, since output can efficiently be computed by combinations of the integrals $E(\alpha,\beta)$ over suitable triangles in the Preisach plane [23], the so-called *Everett integrals*. Nevertheless, in some cases (*e.g.*, in dynamic models), the availability of the distribution function can be very useful. The relationship between the Everett integrals and the PDF is

$$\mu(\alpha,\beta) = -\frac{\partial^2 E(\alpha,\beta)}{\partial\alpha\partial\beta}. \tag{6.2}$$

The identification procedure proposed by the authors is alternative to the classical one and is based on the reconstruction of the Everett function *via* feed forward neural networks (FF-NN) or *via* fuzzy approximators. Thanks to the filtering capabilities of the fuzzy and Feed Forward interpolators [34] no special smoothing of the measured data is necessary for identification.

The first step of both procedures is the building of the training set, constituted by the samples of $E(\alpha,\beta)$ measured on the first-order reversal curves. The training set is then normalised to improve the performance of the subsequent optimisation phase. The second step in the fuzzy procedure is to define the functional expression of the fuzzy interpolator, *i.e.*, of the functional $\tilde{E}(\alpha,\beta,\Theta)$, where $\Theta \in \mathbb{R}^n$ is the vector of the n unknown parameters, which has to approximate $E(\alpha,\beta)$. By referring to the structure in [34], the functional can be written in the form

$$\tilde{E}(\alpha,\beta,\Theta) = \frac{\sum_{i=1}^{n_R} c_i \gamma_{A_{\alpha i}}(\alpha)\gamma_{A_{\beta i}}(\beta)}{\sum_{i=1}^{n_R} \gamma_{A_{\alpha i}}(\alpha)\gamma_{A_{\beta i}}(\beta)}, \tag{6.3}$$

where $\gamma_{A_{\alpha i}}, \gamma_{A_{\beta i}}$ are double-sided Gaussians membership functions of fuzzy sets, whose parameters, together with the weights c_i constitute the vector Θ of the parameters. They can be determined by resorting to an optimisation procedure, whose target is to minimise the cost function

$$f(\Theta) = \log\left(1 + 0.5\sum_{k=1}^{N}[\tilde{E}(\alpha_k,\beta_k,\Theta) - E_k]^2\right), \tag{6.4}$$

where N is the dimension of the training set containing the elements (α_k,β_k,E_k), $k = 1,\ldots,N$.

It can be shown that for any $\delta > 0$ there exist a positive n^* and a parameter vector $\Theta^* \in \mathbb{R}^{n^*}$ such that for all $(\alpha,\beta) \in T$, where T is, in this case, the Preisach triangle, the maximum error satisfies (see [34] and references therein)

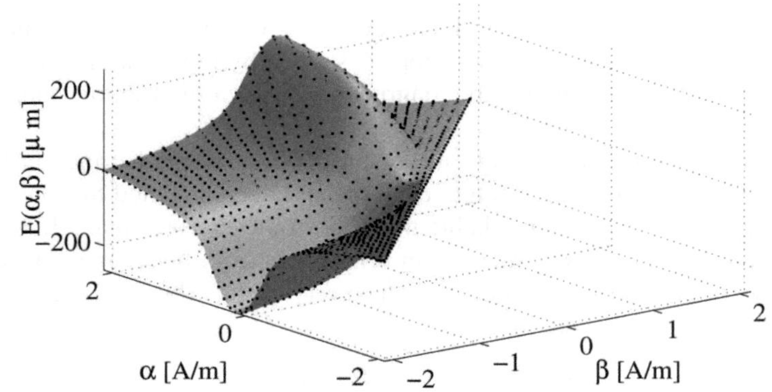

Figure 6.4 Everett integrals for a magnetostrictive actuator: measured data (*dots*) and surface of the identified analytical function

$$\max_{(\alpha,\beta)\in T} \left| E(\alpha,\beta) - \tilde{E}(\alpha,\beta,\Theta^*) \right| \leq \delta. \tag{6.5}$$

Figure 6.4 shows the reconstruction of Everett integrals for a real magnetostrictive actuator. More details and figures of identified distribution functions for different magnetostrictive transducers can be found in [26].

6.4.2 Hysteresis Compensation: Methodology

The concept of inverse (or compensator) of a given hysteretic transducer has been tackled by several authors (see, *e.g.*, [19, 7, 20, 33]). A simplified version of Krasnoselskii's definition is as follows. A system Z is called the compensator (or inverse) of a system Y if, for any initial state y_0 of the system Y, there is a state z_0 of the system Z such that, starting from the states y_0 and z_0, the series connection of the two systems gives the identity transformation independently of their order.

The classical PM (6.1) has an exact inverse if the distribution function $\mu(\alpha,\beta)$ is non-negative [6]. Only in few simple cases (*e.g.*, very simple Preisach distribution functions) is it possible to give a specific mathematical expression of the inverse operator [20], which in turn allows one to tackle several interesting technical problems [3, 4]. In other applications, it is necessary to have more general Preisach models for which an exact expression of the compensator is unavailable. It, therefore, raises the need to perform numerically the compensation of the operator (which can be cumbersome) or to build up a suitable and numerically feasible approximation of it.

The idea of pseudo-compensator as the approximate version of the inverse of a Preisach hysteresis model has been tackled in [26] and is based on the following observation. Since PM fulfills congruency, its compensator is not a PM. Nevertheless, an hysteresis operator of Preisach type $\tilde{\Gamma}$ can be built up such that the condition $\tilde{\Gamma} \circ \Gamma = I$ holds on the first-order reversal branches. Since this transducer cannot be a PM, a nonzero error in the compensation is expected on higher-order reversals. The $\tilde{\Gamma}$ operator is defined as

$$x(t) \equiv \tilde{\Gamma}[y](t) = \iint_{\alpha \geq \beta} \nu(\alpha, \beta)\, \widehat{\gamma}_{\alpha\beta}\, y(t)\, \mathrm{d}\alpha \mathrm{d}\beta, \tag{6.6}$$

where the role of input and output with respect to Equation 6.1 has been interchanged. The identification procedure [26], briefly recalled in Section 6.4.1, allows one to reconstruct the ν distribution function that guarantees, on the first-order reversals, the condition specified above.

Another well-assessed technique to model hysteretic behaviour is based on the modified Prandtl–Ishlinskii hysteresis operator that is represented by the following equation:

$$\Gamma[x] = S[H[x]], \tag{6.7}$$

where H is a Prandtl–Ishlinskii hysteresis operator and S is a Prandtl–Ishlinskii superposition operator. The parameter identification of this model can be formulated as a quadratic optimisation problem that produces the best L_2^2-norm approximation for the measured input–output data of the real hysteretic nonlinearity. Special linear inequality and equality constraints for the parameters guarantee the unique solvability of the identification problem and the invertibility of the identified model. This leads to a robustness of the identification procedure against unknown measurement errors, unknown model errors and unknown model orders. The corresponding compensator can be directly calculated and thus efficiently implemented from the model by analytical transformation laws. All the details of this technique are reported in [21, 22].

6.4.3 Hysteresis Compensation: Experimental Results

The identification of the pseudo-compensator has been experimentally applied to a magnetostrictive linear motor, produced by Energen Inc., model 4-Rxx, equipped with displacement and current sensors. The actuator relation from input current to measured displacement is reported in Figure 6.5. The resulting pseudo-compensator has been implemented in a real-time prototyping system and connected to the real actuator as shown in Figure 6.6. In this way, the compensated actuator has the desired displacement as input and the pseudo-compensator computes the driving current for the magnetostrictive motor. The input signal was a sinusoidal profile with decaying amplitude, so as to generate high-order hysteresis branches. Compensation is very good, as Figure 6.7 shows, despite the fact that the pseudo-compensator is not the exact inverse of the assigned Preisach model. The advantages of using the compensated actuator in a micropositioning and a force control application are described and discussed in [11] and in [12], respectively.

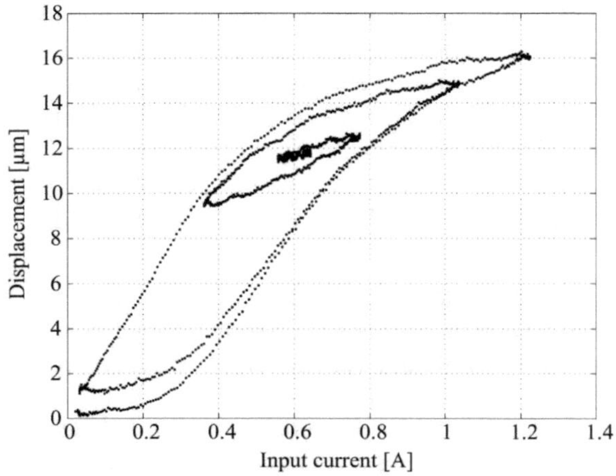

Figure 6.5 Current-displacement relation for the magnetostrictive actuator

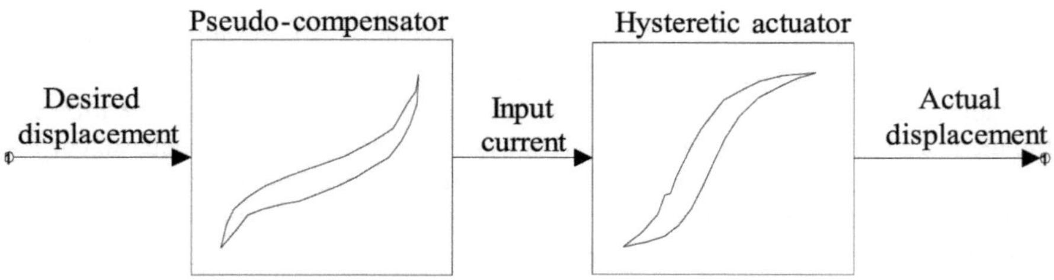

Figure 6.6 Scheme of the actuator hysteresis compensation

6.5 A Smart Auxiliary Mass Damper

In recent years, advanced applications have increasingly required high performance control systems in order to satisfy demanding specifications. These systems often need smart transducers that are specifically designed, although highly nonlinear. As discussed above, smart actuators on the one hand offer a high potential while, on the other hand, may complicate control design preventing the use of linear techniques. This section presents some results on innovative magnetostrictive actuators specially designed to implement an active broadband vibration and noise control system on a full-scale test rig consisting of a segment of a civil aircraft fuselage. The problem addressed has been the reduction of cabin noise caused by turbulence induced vibrations on the fuselage exciting the skin panels. The typical bump noise is found in a broad frequency band ranging from 100 to 400 Hz, depending on the value of the trim velocity of the aircraft. After a preliminary structural and acoustic analysis of a test article the actuator specifications were derived in terms of required forces and maximum allowed weight. In detail, each actuator should be able to generate a force spectrum with a peak of 6 N at 130 Hz with a maximum weight of 150 g.

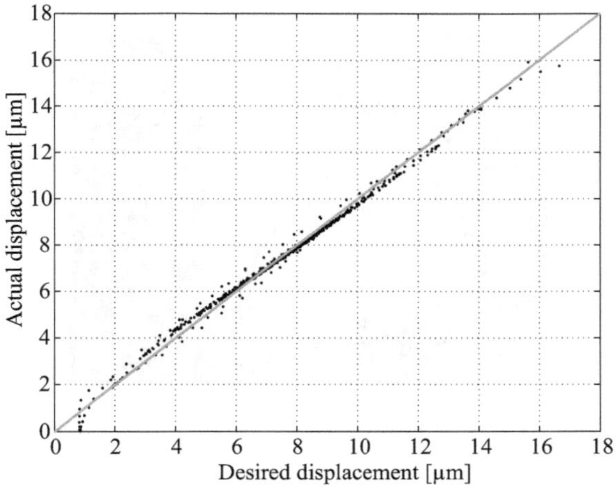

Figure 6.7 Characteristic of the magnetostrictive motor with hysteresis compensation

A conventional actuator could never meet such demanding requirements. Therefore, based on a proven and patented actuator concept [24], the magnetostrictive auxiliary mass damper, shown in Figure 6.8, was optimally designed for this application [25]. The dynamic behaviour of the smart auxiliary mass damper depends strongly on two sources of nonlinearity. The first one is due to the elastic suspension kinematics used for displacement amplification; the second one is due to the hysteretic behaviour of the magnetostrictive material. As reported in the previous section, the latter nonlinear effect can be compensated to improve the performance of the smart auxiliary mass damper when the actuator is used in a feedback control system. The compensation for the described application has been successfully accomplished by means of an hysteresis inversion algorithm [21, 22] implemented on an FPGA based hardware platform as detailed in [16]. This section discusses the problems related only to the kinematic nonlinearity that results in a shift of the actuator's resonant frequency with respect to the driving current amplitude. This nonlinear dynamic behaviour not only complicates the use of the actuator in the control system but also prevents the actuator from delivering the required force at the desired frequency. So it is fundamental to implement a low-level control algorithm that is able to fix the actuator resonant frequency.

The selected approach to design the low-level controller is based on a model-following control strategy [2], which uses the measured displacement of the actuators seismic mass. The main characteristic of the model-following algorithm consists in preserving the nature of the input signal to the low-level control system, which is the output of the high-level noise and vibration control system. Measurements of the actuator seismic mass displacement were carried out by use of a fiber Bragg grating (FBG) sensor. FBG sensors exhibit a number of attractive features such as high sensitivity, lightweight, small size and immunity to electromagnetic interference [17]. The latter point was especially important for this kind of applica-

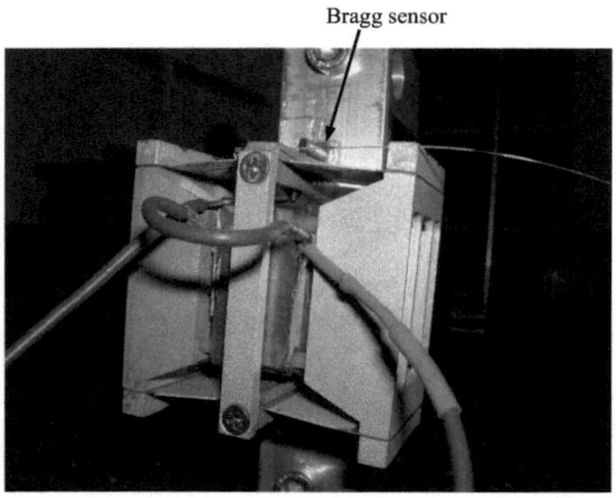

Figure 6.8 Smart auxiliary mass damper

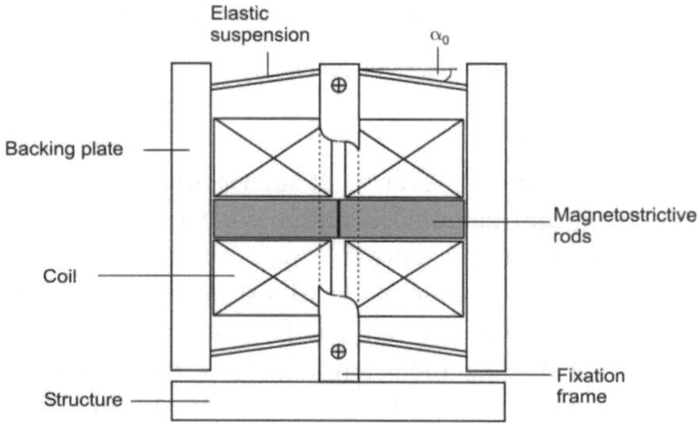

Figure 6.9 Schematic design of the smart auxiliary mass damper

tion, as the magnetostrictive nature of the actuator would be a source of disturbances in electric-based or magnetic-based sensors. The effectiveness of the proposed control strategy has been demonstrated through experiments whose results show that the controlled auxiliary mass damper has a resonance fixed at the desired frequency value. Moreover, the controlled actuator has been tested in an advanced system for vibration control of flexible structures; the details are reported in Chapter 7.

6.5.1 Actuator Description and Modelling

Figure 6.9 shows the fundamental construction inside the smart auxiliary mass damper [24]. It consists of magnetostrictive rods surrounded by two coils. The coils are on two backing plates that are connected with the stiff frame *via* two elastic suspensions arranged in parallel. The frame itself is mounted to the vibrating mechanical structure. A pre-load is applied to the magnetostrictive rods *via* a mechanical

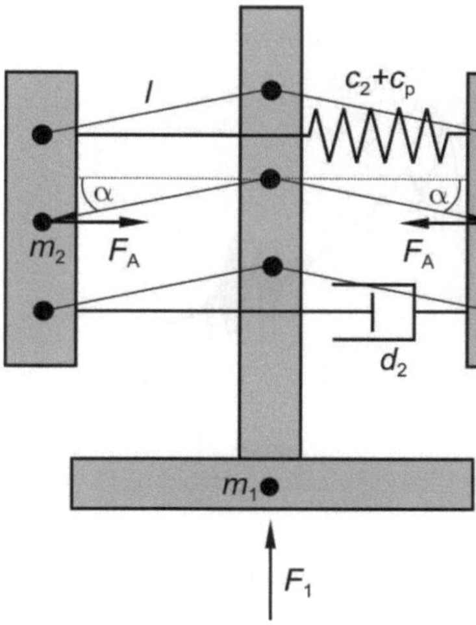

Figure 6.10 Schematic diagram of the actuator mechanics

spring positioned on the top side of the actuator. Due to the magnetostrictive effect a magnetic field caused by a driving current $I(t)$ in the coils produces a small extension $s_A(t)$ in the magnetostrictive rods in the horizontal direction. This extension is transformed by the elastic suspensions to a significantly larger motion of the total mass – consisting of the magnetostrictive rods, the coils and the backing plates – in the perpendicular direction. The displacement amplification strongly depends on the angle of the elastic suspension at the working point α_0 of the mechanical construction (see Figure 6.9) and is larger the smaller this angle α_0 is chosen. As a result of Newton's second law, the total moved mass produces an inertial force that has an effect on the vibrating mechanical structure [24, 25, 27, 28].

The displacement amplification is represented in a first approximation by two parallel rigid arms of length l suspended at play-free and frictionless joints (see Figure 6.10).

The elasticity of the suspension and the mechanical pre-load spring for the actuator are modelled through the linear elastic spring with stiffness c_2. The structural damping in this construction is described by the linear viscous damper with damping d_2. The masses of the backing plates, coils as well as of the moving portions of the suspension and the active material are concentrated in the two identical masses m_2, so that the moved auxiliary mass consists of $2m_2$. The mass m_1 describes the effective mass of the structure to be dampened. The spring with stiffness c_p and the active force F_A describe the passive mechanical behaviour of the active component. The force F_1 represents the unknown disturbance at the point of the structure whose oscillations are to be reduced.

The state variables are the suspension angle $\alpha(t)$ and the angular velocity $\omega(t)$. A nonlinear mathematical model in state space was derived according to the Lagrange formalism [29]. The state space equations are

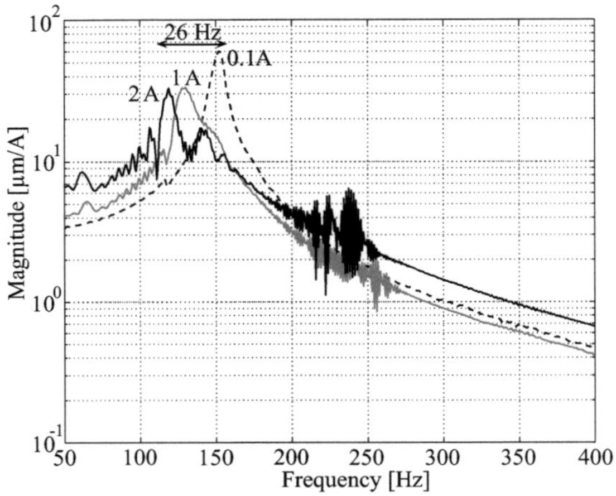

Figure 6.11 Magnitude of frequency response function of the actuator for different current amplitudes

$$\begin{aligned}\dot{\omega} &= -n_1(\alpha)\omega^2 - n_2(\alpha)\omega - n_3(\alpha,\alpha_0)\\ &\quad - n_4(\alpha)s_A + n_5(\alpha)F_1, \qquad (6.8)\\ \dot{\alpha} &= \omega, \qquad (6.9)\end{aligned}$$

with

$$n_1(\alpha) = \frac{2m_2 \sin\alpha\cos\alpha}{m_1 + 2m_2\sin^2\alpha}, \qquad (6.10)$$

$$n_2(\alpha) = \frac{2d_2(m_1+2m_2)\sin^2\alpha}{m_2(m_1+2m_2\sin^2\alpha)}, \qquad (6.11)$$

$$n_3(\alpha,\alpha_0) = \frac{2(c_2+c_p)(m_1+2m_2)(\cos\alpha_0 - \cos\alpha)\sin\alpha}{m_2(m_1+2m_2\sin^2\alpha)}, \qquad (6.12)$$

$$n_4(\alpha) = \frac{c_p(m_1+2m_2)\sin\alpha}{lm_2(m_1+2m_2\sin^2\alpha)}, \qquad (6.13)$$

$$n_5(\alpha) = \frac{\cos\alpha}{l(m_1+2m_2\sin^2\alpha)}. \qquad (6.14)$$

This nonlinear model is able to reproduce the typical nonlinear effect of resonant frequency shifting with respect to input current amplitude, experimentally detected as shown in Figure 6.11.

6.5.2 The Optical Sensor

The possibility to use a FBG as strain sensor is related to the changes in its optical reflection spectrum produced by an applied strain [17]. In particular, the centre wavelength of the FBG reflection spectrum is linearly dependent on strain, so the latter can be retrieved on a wavelength-encoding basis. The dynamic strain measurements were carried out by employing a narrow-band demodulation scheme. The output light wave from a single longitudinal mode distributed feedback (DFB) diode laser was used to probe the wavelength shift of the FBG reflection curve imposed by the strain signal to be detected. If the laser frequency is within the linear range of the FBG reflection slope, the strain signals will change the reflected power, which can be simply measured using a photodetector [9]. It can be seen that this technique has several advantages, such as low cost, fast response and ease of use. The reflectivity spectrum of the used FBG used had a quasi-flat top from 1552 to 1556 nm, with a peak reflectivity of $\geq 75\%$ (see Figure 6.12). The leading edge of the FBG reflection curve extended from 1547.28 to 1550.95 nm (measured from 10% to 90% of maximum reflectivity), whereas the trailing edge extended from 1556.37 to 1557.24 nm. The trailing edge of the FBG reflection curve allowed for a 72 GHz linear slope width and it was chosen as the operating range due to the higher linearity exhibited by the FBG reflectivity in this spectral portion. In order to keep a linear relationship between the reflected optical power and the strain signal, the DFB laser emitting wavelength must lie within the trailing edge of the FBG reflectivity spectrum (see Figure 6.12). Moreover, assuming a typical FBG curve shift of 1 nm for an applied strain of 1000 $\mu\varepsilon$, the strain level must be kept lower than 870 $\mu\varepsilon$ in order to avoid sensor output saturation.

The optical circuit employed for FBG interrogation is schematically illustrated in Figure 6.13. Light emitted by a DFB laser was sent to a Y-coupler, which directed the light to the FBG. Light reflected from the FBG was then re-directed to a photodiode, whose bandwidth was of the order of hundreds MHz, thus much higher than the bandwidth of the strain signals to be detected. The output signal from the photodiode was finally sent to the conditioning electronics, basically comprised of a bandpass filter used to eliminate the dc component and high frequency noise.

Before attaching the FBG to the magnetostrictive actuator, the portion of fibre jacket corresponding to the grating position was removed, so as to avoid strain transfer loss from actuator to FBG. The FBG was then pre-stressed and glued to the actuator by epoxy resin (see Figure 6.8). Pre-stressing of the FBG was necessary in order to avoid buckling of the fibre in which the sensor is written. Mounting of the FBG was carried out with the aim of measuring one of the state variables of the system $\alpha(t)$. A sketch of the mounting setup is shown in Figure 6.14. As the output provided by the FBG is proportional to the strain of the fibre segment between the two bonding points A and B, the relationship between the strain and the angular displacement must be determined. Such a relationship can be obtained by means of geometrical considerations. Referring to Figure 6.14, start by considering that, under real experimental conditions, the displacement Δx of the moving mass is much smaller than the lever arm length l, *i.e.*, $\Delta x \approx l\Delta\alpha$. Moreover, Δx is also

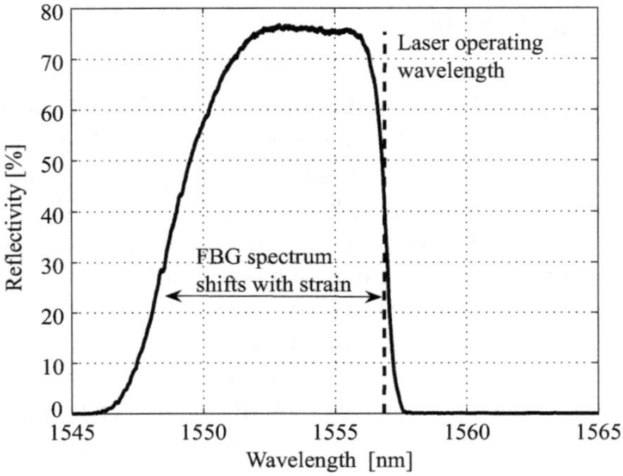

Figure 6.12 Reflectivity spectrum of the grating

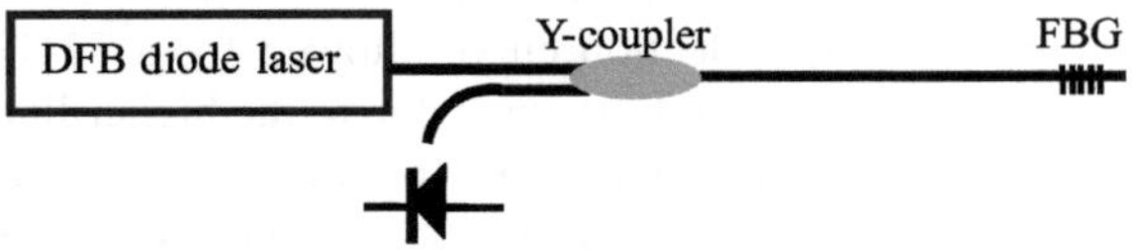

Figure 6.13 FBG interrogation scheme

much smaller than the fibre segment b, so that $\Delta\beta \approx 0$ can also be assumed. Under this approximation, the displacement Δx is nearly equal to Δb. As the optical output is proportional to the strain $\Delta b/b$, it is possible to write

$$\varepsilon \equiv \frac{\Delta b}{b} \approx \frac{l}{b}\Delta\alpha. \tag{6.15}$$

Hence, under the small signal hypothesis, the quantity measured by the FBG is directly proportional to the angle variation $\Delta\alpha = \alpha - \alpha_0$. After bonding the FBG, the DFB laser output wavelength was temperature-tuned in order to guarantee a linear response of the sensor, over the whole range of actuator displacement. Calibration of the FBG was carried out by aid of an accelerometer mounted to the moving mass, so as to measure the acceleration along the vertical axis. The sensor sensitivity was estimated by comparing the measured frequency response functions from the excitation signal to the second time derivative of the FBG output signal, and from the excitation signal to the accelerometer output signal, respectively. Results are shown in Figure 6.15. Based on the calibration constant of the accelerometer, an FBG sensitivity of about 37.6 mV/μm has been estimated.

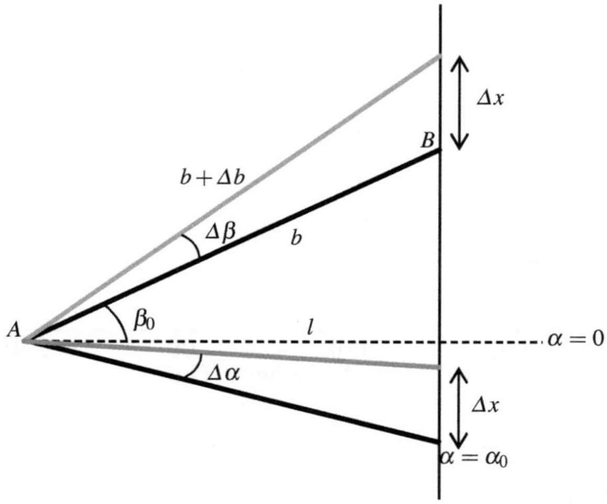

Figure 6.14 Sketch of mounting setup

6.5.3 The Control System

As mentioned before, the actuator has a resonant frequency that shifts with respect to the driving current amplitude. This nonlinearity makes the actuator difficult to implement in the active vibration control system. Therefore, the actuator control objective is to keep constant the resonance frequency of the actuator despite the amplitude of the input current.

First of all, a linear model reproducing the behaviour of the actuator in the neighbourhood of a given working condition, *i.e.*, with a given current amplitude and a fixed resonant frequency, is identified using a frequency domain identification procedure. The model can be written in the classical state space form as

$$\dot{x}_a(t) = Ax_a(t) + Bu(t), \tag{6.16}$$

$$y_a(t) = Cx_a(t) + Du(t), \tag{6.17}$$

where $u(t)$ is the actuator input current and $y_a(t)$ is the measured FBG signal. The state space Equations 6.16 and 6.17 correspond to the linearised version of the nonlinear model of Equations 6.8 and 6.9, around the initial suspension angle α_0.

The adopted control strategy is based on a model-following approach [2] and the control scheme is reported in Figure 6.16. The characteristic of the model-following algorithm consists of preserving the nature of the input signal of the control system. This makes the use of the actuator more transparent in the higher level control system computing the reference current for the actuator as a result of an outer feedback loop. This is accomplished by defining a reference model whose input is just the reference current $u_r(t)$, while the actuator input is the sum of the reference current and a corrective current $u_c(t)$ computed by the controller on the basis of the output error

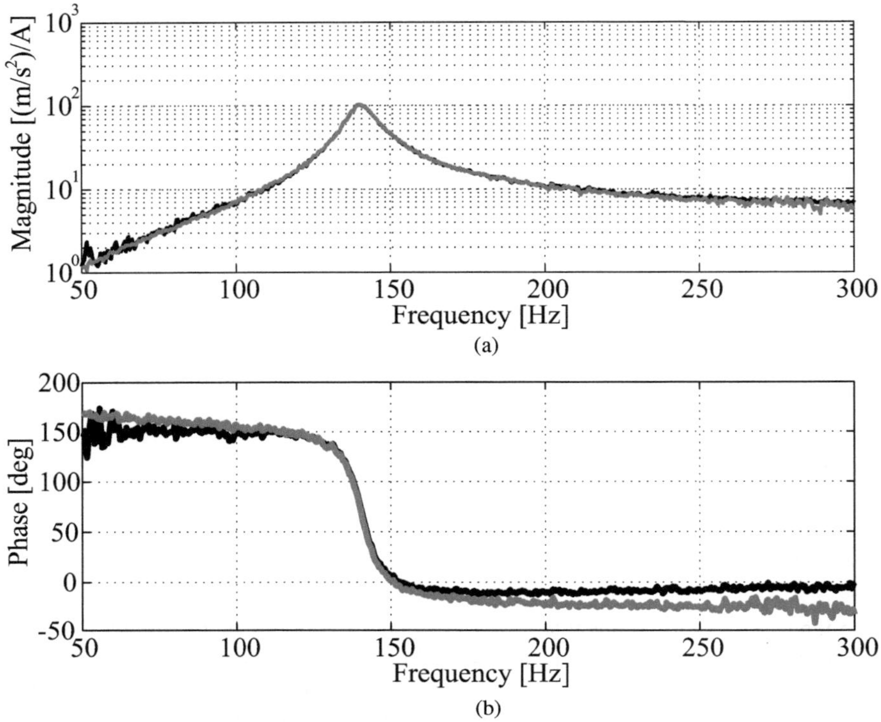

Figure 6.15 (a) Magnitude and (b) phase of the FRF for sensor calibration: accelerometer output (*solid black lines*), second time derivative of the FBG output signal (*solid grey lines*)

$e_y(t) = y_a(t) - y_r(t)$, $y_r(t)$ being the reference output. In particular, the reference model is selected equal to the identified linear model of the actuator, *i.e.*,

$$\dot{x}_r(t) = Ax_r(t) + Bu_r(t), \tag{6.18}$$

$$y_r(t) = Cx_r(t) + Du_r(t). \tag{6.19}$$

As a consequence, it is easy to verify that the dynamics of the state error $e_x(t) = x_a(t) - x_r(t)$ is described by the state space representation

$$\dot{e}_x(t) = Ae_x(t) + Bu_c(t), \tag{6.20}$$

$$e_y(t) = Ce_x(t) + Du_c(t). \tag{6.21}$$

Since the state e_x is obviously not accessible, the controller is designed according to a standard LQG procedure consisting of a Kalman filter as observer of the error dynamics and an optimal state feedback regulator. Therefore, its dynamic equations can be written in the form

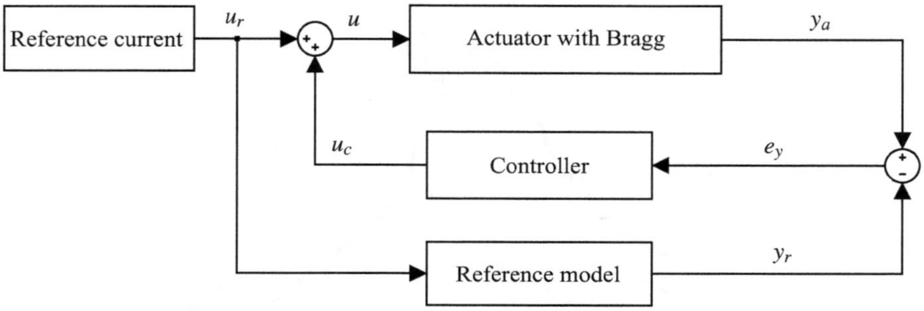

Figure 6.16 Reference model control scheme

$$\dot{x}_c(t) = A_c x_c(t) + B_c e_y(t), \tag{6.22}$$

$$u_c(t) = C_c x_c(t), \tag{6.23}$$

where $A_c = A - LC - BK + LDK$, $B_c = L$ and $C_c = -K$, with L and K being the solutions of two classical algebraic Riccati equations.

6.5.4 Experimental Results of Actuator Control

The identification of the reference model has been carried out using FDIDENT, a third-party MATLAB toolbox for frequency domain model identification and validation, based on the algorithm originally presented in [18]. The actuator was excited with a chirp input current in the frequency range $[50, 600]$ Hz with amplitude 1 A, repeated eight times, while acquiring the signal from the FBG sensor. Then, the frequency response function mapping the current to the FBG signal was estimated with eight averages to reduce noise effect and has been used as input data for the identification procedure. The reference model resulting from the identification is a sixth-order model.

The controller was designed solving the LQG problem, with a process noise covariance matrix

$$Q = q_0 I + q B_r B_r^{\mathrm{T}}, \tag{6.24}$$

with $q_0 = 10^{-4}$, I the identity matrix, $q = 0.5 \cdot 10^{-2}$ and as measurement noise covariance $v = 0.2 \cdot 10^{-3}$. The weights for the regulator problem were chosen as $W = C^{\mathrm{T}} C$ for the state weight and $r = 0.1$ for the input weight. Figure 6.17 reports the effect of the implemented controller on the resonant frequency of the actuator. It is evident that the resonance remains fixed even for different input current amplitudes when the closed loop control is active.

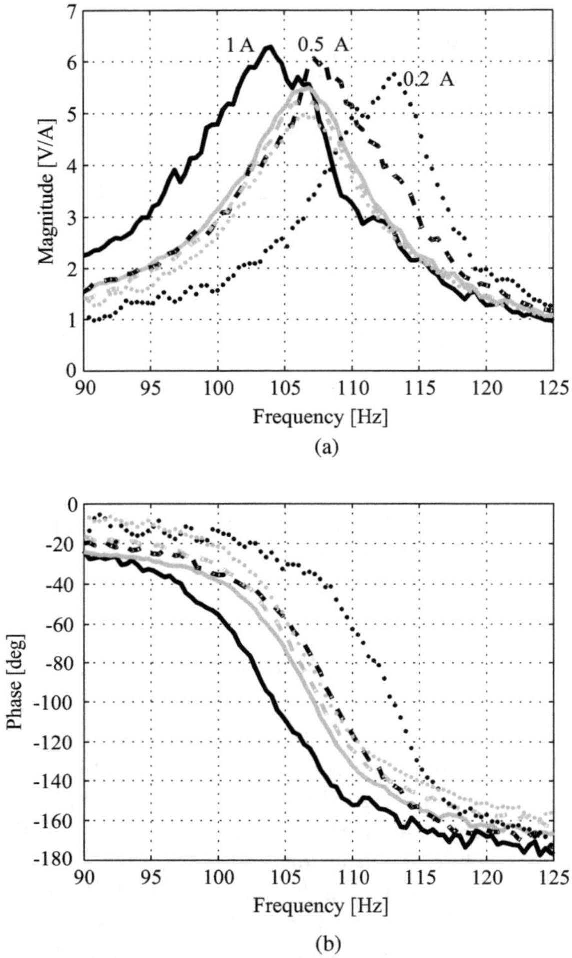

Figure 6.17 (a) Magnitude and (b) phase of the FRF of the actuator for different current amplitudes: open loop (*black lines*) and closed loop (*grey lines*)

6.6 Self-sensing Control

The importance of sensor-actuator colocation in active vibration control was pointed out in Chapter 5. The piezoelectric and magnetostrictive devices can be used in the so-called "self-sensing" configuration, *i.e.*, the same transducer is used simultaneously as actuator and as sensor. This configuration guarantees that the actuated system is minimum phase. The self-sensing configuration has been widely used in experiments of colocated control [1, 14, 32]. This section describes the implementation of the self-sensing circuit for a piezoelectric patch. As pointed out in [14], the main difficulty in effectively using a piezoelectric device in self-sensing con-

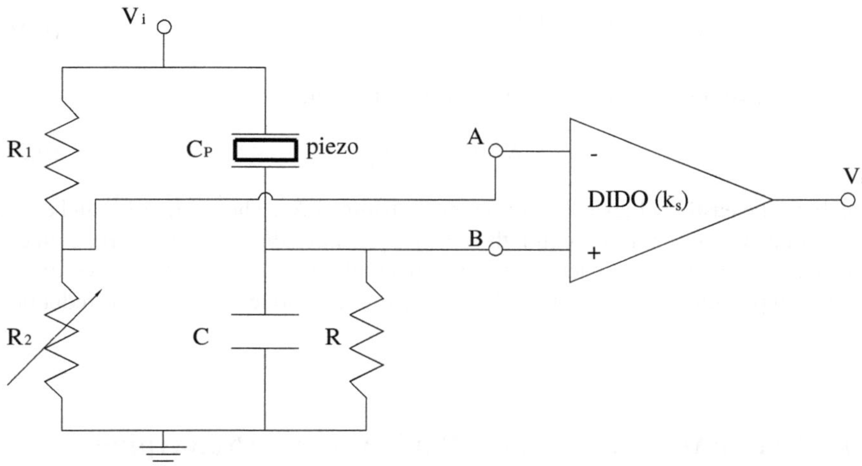

Figure 6.18 Schematic of the self-sensing circuit

figuration is represented by the tuning of components' values in the typical bridge structure of the circuit. In the above literature, a self-sensing bridge is usually devised so that the piezo is connected to the first branch, while in the other branch a capacitor is inserted so as to balance the impedance of the piezo capacitance. This method is affected by the difficulty of tuning the value of balancing capacitance because of the limited accuracy of variable capacitors and variation of the capacitance values with time. In order to overcome this drawback a modification to the self-sensing circuit is proposed to facilitate the balancing of the bridge, even though the balancing is theoretically obtainable only in a frequency range that, on the other hand, is practically very wide.

The schematic of the circuit is depicted in Figure 6.18. It can be seen how the capacitive branch has been replaced by a pure resistive voltage divider, so that a very accurate trimmer resistor can be used to balance the bridge. By assuming for the differential input differential output (DIDO) amplifier an infinite input resistance, the sensed voltage, *i.e.*, the output of the DIDO amplifier is given by

$$V_s = k_s \left[V_i \left(\frac{1}{1+Z_p/Z} - \frac{1}{1+R_1/R_2} \right) + \Delta V \frac{1}{1+Z_p/Z} \right], \tag{6.25}$$

where Z and Z_p are the impedances on the right branch of the bridge, k_s is the gain of the DIDO amplifier and ΔV is the voltage generated by the piezo device according to the piezoelectric effect, namely, the voltage related to the strain of the structure in the point where the piezo is bonded [14]. It is evident that balancing cannot be obtained for every frequency, however, if the minimum frequency of interest f_m for the signal $\Delta V(t)$ is such that $2\pi f_m >> 1/RC$, then the following approximation can be made:

$$\frac{Z_p}{Z} = \frac{1+sRC}{sRC_p} \simeq \frac{C_p}{C}, \quad \forall f \geq f_m. \tag{6.26}$$

Under this assumption, the balancing equation becomes

$$R_1/R_2 = C/C_p, \tag{6.27}$$

which can be easily achieved by acting on the trimmer R_2. The values of R and C can be designed by taking into account the necessity to have the most of V_i falling on C_p, to properly drive the piezo actuator, hence it should be $C >> C_p$. A detailed experimental application of the proposed self-sensing technique for vibration reduction can found in [8].

6.7 The dSPACE Platform for Rapid Control Prototyping

Rapid control prototyping (RCP) is a process that allows the engineers to quickly test and iterate their control strategies on a real-time system connected to real input–output devices. In rapid control prototyping, differently from hardware-in-the-loop, the control strategy is simulated in real-time and the system under control is the actual one. RCP is nowadays the typical method used by engineers to develop and test their control strategies.

RCP had a fast development in the automotive market in the mid 1990s. It was a welcome solution to managing the increasing complexity of the control tasks. Since then the use of rapid control prototyping has grown in other fields such as the aerospace industry, medical device development and industrial automation applications. Furthermore, low cost RCP systems are being used as teaching tools in most university graduate-level courses in classical and modern control theory.

The typical RCP system is comprised of the following components:

1. A math modelling software such as MATLAB/SIMULINK with a real-time development environment as such MATLAB Real-Time Workshop.
2. A real-time target system that is typically an embedded computer with input–output interfaces such as analogue, digital and serial.
3. A host PC with communications link to the target system.
4. A graphical user interface (GUI) application to download and control the real-time process from the host PC.

One of the reasons that has made RCP so popular is that the algorithms can be developed directly with graphical models using math modelling packages such as MATLAB/SIMULINK rather than programming languages such as C-code. Engineers can concentrate on developing control strategies in a familiar modelling environment and do not have to worry about translating the models to other programming codes suitable for the real-time target. The rapid control prototyping system does this work using an automated process. From a practical point of view, a typical development procedure includes the following steps:

1. The control strategy is developed using a graphical math modelling application such as Simulink.
2. Symbolic I/O blocks, available depending on the used real-time hardware, are imported into the math model and connected to appropriate points in order to add input–output capabilities to the control strategy.
3. The "build" process is invoked. This process reads the model and uses an automatic code generator to translate the model in C-code. Then the code is compiled and linked with target specific code such as a scheduler, I/O routines and communication routines specific to the real-time target. The end result is an executable program for the target real-time computer.
4. Using a GUI program, installed on the host PC, the executable program is downloaded to the target processor and the program can be controlled and instrumented by the GUI program. Variables can be monitored, graphed or logged. The strategy can be altered by only tuning gains, while the structure of the strategy cannot be changed.

A leading company in the area of RCP commercial systems is dSPACE GmbH. The dSPACE prototyping systems are flexible development systems that enable development and optimisation of control designs for the real plant without textual programming. With dSPACE implementation software Real-Time Interface (RTI), MATLAB/Simulink models can be implemented on the hardware automatically. A graphical Simulink block library with numerous interface functionalities is provided for configuring inputs and outputs and connecting I/O to the model. The scalable, modular and highly configurable hardware for signal conditioning and power stages allows direct connection of the dSPACE prototyping hardware to sensors and actuators. Moreover, dSPACE has a wide range of off-the-shelf software and hardware components for building tailored prototyping systems. All the experiments reported in this book were performed using dSPACE modular systems whose specifications are described below.

6.7.1 Processor Boards

Processor boards are the basis of dSPACE modular hardware. They give to the real-time system computing power and also provide the interfaces to the I/O boards *via* a fast 32-bit I/O bus (PHS bus), and to the host PC *via* the PCI bus. The processor boards also do all the communication with experiment tools running on the host computer. There might be some cases where a single processor system is not sufficient, for example, when the application requires extremely large real-time models or more than 16 I/O boards. It is possible to reach any performance and I/O capacity by simply expanding the hardware with further processor boards. The processor boards are connected *via* Gigalink modules, which allow high-speed serial data transmission owing to a fibre-optic cable and a proprietary transmission protocol, leading to a 1.25 Gbit/s communication speed over distances exceeding 100 m. One of the processor boards for dSPACE modular hardware is the DS1005 PPC board. This board is the heart of many dSPACE systems for rapid control prototyping and is equipped with the fastest available PowerPC 750GX processor running at 1 GHz from IBM. Processing power and I/O capacity can be boosted even further by linking up to 20 DS1005 PPC Boards in a multiprocessor system. The DS1005 is directly connectable to all dSPACE I/O boards *via* PHS. Great processor power plus fast access to I/O hardware with minimum latencies make dSPACE processor boards considerably faster than solutions based on commonly available PCs. The modular systems used for the experiments described in the book are equipped with a single DS1005 board or with two processor boards connected in parallel.

6.7.2 I/O Boards

Modular hardware usually contains at least one I/O board for connection to the real world. To meet all the demands of rapid control prototyping, dSPACE systems offer a large number of I/O boards with different kind of channel. Analog-to-digital (A/D) and digital-to-analog (D/A) converters with different resolutions and speeds are available. Many boards include digital I/O such as pure bit channels, control switches, displays and support timing functions such as the generation and capture of PWM or square-wave signals, and also other pulse patterns. Also more specialised I/O like CAN, LIN, FlexRay and incremental encoders can be easily obtained. Many boards are multifunctional since they offer both digital I/O and A/D and D/A channels. The fundamental element of most applications in rapid control prototyping is the A/D and D/A conversion, since sensors and actuators employed in feedback control systems produce or need analogue voltages or currents. It is important to select a conversion board with channel numbers, resolution, speed and voltage range appropriate to the experiment. The right selection of these parameters avoids implementation problems associated with the signals digitalisation. The A/D boards used in the experiments presented in this book are of two types: the DS2001 ADC board that has five independent channels each with a 16-bit maximum resolution converter and a sample time of 5 μs; the DS2004 ADC high-speed board that has 16 independent channels (also working in differential mode) each with a 16-bit resolution converter and a sample time of 800 ns. Both boards can operate with ± 5 V or ± 10 V as input voltage range. However the D/A conversion was always performed with DS2103 DAC board that integrates 32 parallel D/A converters each with a resolution of 14-bit and a settling time of 10 μs. The output voltage range is programmable to ± 5 V or ± 10 V.

6.7.3 Software Development Environment

The implementation software for running models on dSPACE hardware is RTI that is the link between dSPACE systems and the development software MAT-

LAB/Simulink. It extends Real-Time Workshop (C code generator) for the automatic implementation of Simulink models on the real-time hardware. To connect the model to a dSPACE I/O board, it is sufficient to drag and drop the I/O module from the RTI block library and then connect it to the Simulink blocks. All I/O channel settings are available by clicking the appropriate blocks. Real-Time Workshop generates the model code while RTI provides blocks that implement the I/O capabilities of dSPACE systems in the Simulink models. The real-time model is compiled, downloaded and started automatically on the real-time hardware, without writing a single line of code. RTI handles any kind of continuous-time, discrete-time, hybrid and multirate systems. Depending on the I/O hardware, different channels of the same I/O board can be used with different sample rates, and even in different subsystems. RTI supports asynchronous events and allows setting task priorities and task overrun strategies for executing the interrupt-driven subsystems. It also supports time-triggered tasks and timetables, which allow implementing tasks and groups of tasks with variable or predefined delay times in relation to an associated trigger event. This makes task handling in the model very flexible. The multiprocessor option is also available for real-time interface (RTI-MP), which supports multiprocessor systems based on dSPACE modular hardware and increases the performance of real-time simulations. It assists in setting up multiprocessor networks, including communication channels. While the real-time control application is running on the dSPACE hardware the experiment can be managed with the software ControlDesk. It allows building of instrumentation panels for controlling and monitoring the variables of the real-time experiment. The experiment interface can be designed by drag and drop, utilising features such as context menus, numeric inputs, sliding bars, floating windows, displays and many others. Multiple data acquisitions can be synchronised to a reference time with a precision of about 60 μs.

6.8 Conclusions

This chapter provided technical information useful for choosing the type of actuator and/or sensor for the implementation of a vibration control system. Initially, it described the advantages and the disadvantages using smart transducers to satisfy the particularly demanding specifications of aeronautic applications. Previous chapters presented modelling, identification and control techniques for flexible structures under the linearity assumption; therefore this chapter discussed the typical nonlinearity characteristics of smart material-based devices and their effects on the control system. In detail, when these nonlinearities are not negligible, the use of smart transducers to implement linear active control strategies needs an accurate modelling and a suitable control. The chapter presented some techniques that allow proper use of non-conventional devices (typically strongly nonlinear) together with a linear approach for active vibration control. The actuators described together with the rapid control prototype system, presented in the latter section of this chapter, will be used in the experiments illustrated in Chapter 7.

References

1. Anderson EH, Hagood NW (1994) Simultaneous Piezoelectric Sensing/Actuation: Analysis and Application to Controlled Structures. J Sound Vib 174(5):617–639
2. Balestrino A, De Maria G, Zinober ASI (1984) Nonlinear Adaptive Model-following Control. Automatica 20:559–568
3. Basso V (1998) Hysteresis Models for Magnetization by Domain Wall Motion. IEEE Trans Magn 34:2207–2212
4. Basso V, Bertotti G, Serpico C, Visone C (1998) Application of an Exactly Inverse Hysteresis Model to Magnetic Field Computations. J Physique IV 8:639–642
5. Bertotti G (1998) Hysteresis in Magnetism. Academic, New York
6. Brokate M (1989) Some Mathematical Properties of the Preisach Model of Hysteresis. IEEE Trans Magn 25:2922–2924
7. Brokate M, Sprekels J (1996) Hysteresis and Phase Transitions. Springer, New York
8. Cavallo A, De Maria G, Natale C (2001) Second Order Sliding Manifold Approach for Vibration Reduction via Output Feedback: Experimental Results. In: Proc. of 2001 IEEE/ASME International Conference on Advanced Intelligent Mechatronics, Como (I), 725–730
9. Cavallo A, May C, Minardo A, Natale C, Pagliarulo P, Pirozzi S (2009) Active Vibration Control by a Smart Auxiliary Mass Damper Equipped with a Fiber Bragg Grating Sensor. Sens and Act A: Physical 153:180–186
10. Cavallo A, Natale C, Pirozzi S, Visone C, (2005) Limit Cycles in Control Systems Employing Smart Actuators with Hysteresis. IEEE/ASME Trans Mech 10(2):172–180
11. Cavallo A, Natale C, Pirozzi S, Visone C (2003) Effects of Hysteresis Compensation in Feedback Control Systems. IEEE Trans Magn 39:1389–1392
12. Cavallo A, Natale C, Pirozzi S, Visone C (2004) Feedback Control Systems for Micropositioning Tasks with Hysteresis Compensation. IEEE Trans Magn 40:876–879
13. Culshow B (1996) Smart Structures and Materials. Artech House, London
14. Dosch JJ, Inman DJ, Garcia E (1992) A Self-sensing Piezoelectric Actuator for Collocated Control. J Intell Mat Sys Struct 3:166–185
15. Engdahl G (2000) Handbook of Giant Magnetostrictive Materials. Academic, New York
16. Janocha H, Pesotski D, Kuhnen K (2006) FPGA-based Compensator of Hysteretic Actuator Nonlinearities for Highly Dynamic Applications. In: Proc 10th Int Conf New Actuators, Bremen (D), 1013–1016
17. Kersey AD, Davis MA, Patrick HJ, Leblanc M, Koo KP, Askins CG, Putnam MA, Friebele EJ (1997) Fiber Grating Sensors. J Lightw Technol 15:1442–1462
18. Kollár I (1993) On Frequency-Domain Identification of Linear Systems. IEEE Trans Instrum and Meas 42(1):2–6
19. Krasnoselskii MA, Pokrovskii AV (1989) Systems with Hysteresis. Springer, New York
20. Krejci P (1986) Hysteresis and Periodic Solutions of Semilinear and Quasilinear Wave Equations. Mathematische Zeit 193:247–264
21. Kuhnen K (2003) Modelling, Identification and Compensation of Complex Hysteretic Nonlinearities–A Modified Prandtl-Ishlinskii Approach. Eur J Control 9:407–418
22. Kuhnen K, Janocha H (2001) Inverse Feedforward Controller for Complex Hysteretic Nonlinearities in Smart Material Systems. Control Intell Syst 29:74–83
23. Mayergoyz ID (1991) Mathematical Models of Hysteresis. Springer, New York
24. May C, Kuhnen K, Pagliarulo P, Janocha H (2003) Magnetostrictive Dynamic Vibration Absorber (DVA) for Passive and Active Damping. In: Proc 5th Eur Conf Noise Control, Naples (I), 1–6
25. May C, Pagliarulo P, Janocha H (2006) Optimisation of a Magnetostrictive Auxiliary Mass Damper. In: Proc 10th Int Conf New Actuators, Bremen (D), 344–348
26. Natale C, Velardi F, Visone C (2001) Identification and Compensation of Preisach Hysteresis Models for Magnetostrictive Actuators. Physica B 306:161–165
27. Pagliarulo P, Kuhnen K, Janocha H (2007) Adaptronic Vibration Absorber for a Wide Field of Applications. In: Adaptronic Congress, Göttingen (D), 294–303

28. Pagliarulo P, Kuhnen K, May C, Pagliarulo P, Kuhnen K, May C (2004) Tunable Magnetostrictive Dynamic Vibration Absorber. In: Proc 9th Int Conf New Actuators, Bremen (D), 698–701
29. Schultz DG, Meltsa JL (1967) State Functions and Linear Control Systems. McGraw-Hill, New York
30. Schwartz M (2009) Smart Materials. CRC Press, Florida
31. Shahinpoor M, Schneider HJ (2008) Intelligent Materials. RSC Publishing, UK
32. Spangler RL Jr, Hall SR (1994) Broadband Active Structural Damping Using Positive Real Compensation and Piezoelectric Simultaneous Sensing and Actuation. J Smart Mat Struct 3:448-458
33. Visintin A (1991) Differential Models of Hysteresis. Springer, Berlin
34. Wang LX (1992) Fuzzy Systems are Universal Approximators. IEEE Trans Syst Man Cybern 7:1163–1170

Chapter 7
Experimental Results

7.1 Introduction

The identification and control procedures described in previous chapters have been experimentally tested on a very complex flexible structure. In particular, the selected test article is a fuselage skin panel of a BOEING 717 (depicted in Figure 7.1). The panel is stiffened by two bulkheads and three orthogonal stiffeners rivetted to the rear of the panel itself (see Figure 7.2). In order to reproduce the conditions usually adopted in the structural testing phase, the panel has been suspended by a couple of soft springs to simulate the free boundary conditions.

As explained throughout the book, the active vibration control techniques adopted here are all model-based; therefore the first step of the design procedure is to identify the parameters of a dynamic model. This has been accomplished by exploiting the procedure described in Chapter 3, and the results of the experimental identification are reported in the following of this chapter. In particular, the results are used to compare the identified model with other models derived by exploiting different identification algorithms from different point of views. The comparison is performed not only in terms of the model capability to accurately reproduce the system behaviour, but also from the computational point of view as time and skill required from the user, who has to carry out the identification procedure. The main differences among the algorithms are also highlighted in terms of their features, *e.g.*, applicability to MIMO systems, guaranteed structural properties of the obtained model, availability of order reduction technique. The identified model clearly shows that the skin panel is a flexible structure with a high modal density. The model is used to design the active vibration control algorithms according to the approaches described in Chapter 5. In particular, the $\mathscr{H}_2$ and $\mathscr{H}_\infty$ optimal control strategies with only colocated sensor–actuator pairs (see Section 5.1) are implemented using piezo actuators and compared to each other in tackling a problem of vibration reduction when a broadband primary field is applied to the structure. The performance of both controllers is evaluated in terms of indices suitably defined to quantify the vibration reduction on the whole panel. Then, the robust controllers introduced in Section 5.10 requiring an

Figure 7.1 Front view of the panel with the four piezoelectric patches' actuators

LMI-based design procedure are implemented and tested. Firstly, the performance of the controller making use of sole colocated sensor–actuator pairs but designed with different choices of performance weights is evaluated. Secondly, this performance is compared with that of the robust controller exploiting not only colocated sensor–actuator pairs but also additional sensors, suitably placed in the region where the primary field is applied. A third case study is aimed at highlighting the feature of the LMI-based control design that allows the user to concentrate the control effort on a selected group of natural modes of the structure. This is a particularly attractive feature when the knowledge of the frequency range of the primary field spectrum is available and thus the user can concentrate the control effort on the modes whose natural frequencies are within that range. The final case study is devoted to show how the proposed robust control algorithm can be successfully adopted even using only non-colocated sensor–actuator pairs. This feature is extremely important in all those applications when the actuators cannot be placed in the same place where the sensor measures the quantity of interest, *e.g.*, deflection of a beam tip.

Figure 7.2 Rear view of the panel with one bulkhead and two orthogonal stiffeners and with four accelerometers

With the aim of showing that the proposed active control strategies can be applied independently from the particular actuator selected for the application, the last section of the chapter is devoted to presenting the experimental results of active vibration control exploiting the smart auxiliary mass damper described in Chapter 6. In this case study, due to the strong actuator nonlinearities, a suitable low-level control loop has been implemented at actuator level to make possible the application of the same control methodology adopted in the case of piezo actuators.

7.2 Identification of the Dynamic Model of the Panel

The test article shown in Figures 7.1 and 7.2 was equipped with four actuators and four colocated sensors in order to implement the identification procedure described in Chapter 3. In particular, the input forces are applied by means of four piezoelectric patches by ACX bonded on the panel surface; the output acceleration is measured by means of four accelerometers by PCB placed on the rear part of the panel in correspondence with the piezoelectric actuators (see Figure 7.2), so as to achieve the colocation. Further details of the setup and the devices used can be found in the Appendix, where the numerical parameters of the complete model are also reported. The output velocities were estimated by numerical integration in the frequency domain. The frequency range of interest was selected from $f_1 = 100$ to

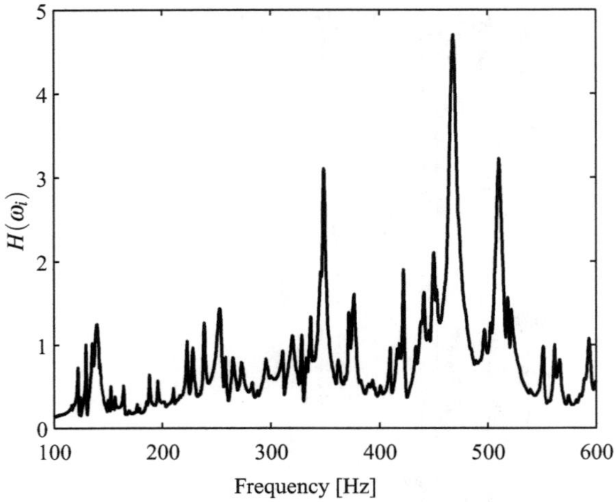

Figure 7.3 Indicator $H(\omega_i)$ defined in Equation 3.44 for the aeronautical panel

$f_2 = 350\,\text{Hz}$. The measurements were performed by using a dSPACE rapid prototyping real-time control system with 16-bit A/D channels and 14-bit D/A channels at a sampling frequency of 20 kHz. The details of the data acquisition system can be found in Chapter 6.

The 4×4 frequency response matrix has been estimated by exciting the system with a chirp signal with a linearly time-varying frequency from $f_1' = 100\,\text{Hz}$ to $f_2' = 600\,\text{Hz}$ and a duration of 15 s. Note that $f_1 = f_1'$ since significant lower frequency modes were discovered during preliminary tests performed on the test article. The input signal duration yields 7500 spectral lines of the acquired signals, 2500 of which are used for identification (defining the set $\mathscr{I} \subset \{1,\ldots,7500\}$ obtained by decimating the initial set by 3) and the remaining for the validation and the optimal selection of the weight α in (3.59). Before starting with the procedure some precautions were taken. Firstly, the accelerometers were carefully placed so that the phase of the colocated FRF starts from $90°$ and belongs to the range $[-90°, 90°]$ as required by the model structure. Then, before estimating the FRF the offset voltages affecting the conditioning electronics of the accelerometers were removed. To minimise the measurement noise, the experiments were repeated four times and the time domain data were suitably composed so as to estimate the FRF by resorting to the Welch's algorithm [10]. Finally, to ensure that the input data exactly satisfy the reciprocity property, a symmetrisation of the estimated FRFs was carried out so that only 10 of the estimated 16 FRFs were actually used.

Before applying the first stage of the procedure originally proposed in [2] , the indicator in Equation 3.44 was computed and plotted (see Figure 7.3), so that the initial number of modes was selected equal to 165 by roughly counting the peaks. The first stage of the procedure leads to the estimation of natural frequencies and damping coefficients of the structure (not reported here for brevity). In the second stage, which computes the input and output matrices, the modal residues are computed

Table 7.1 Values of the quadratic index J_α in Equation 7.1 obtained with different values of α in Equation 3.59

J_α	8.18	6.98	6.39	5.95	5.96	6.00	6.01	6.04	6.08	6.17
α	0.1	0.2	0.3	0.4	0.5	0.6	0.7	0.8	0.9	1.0

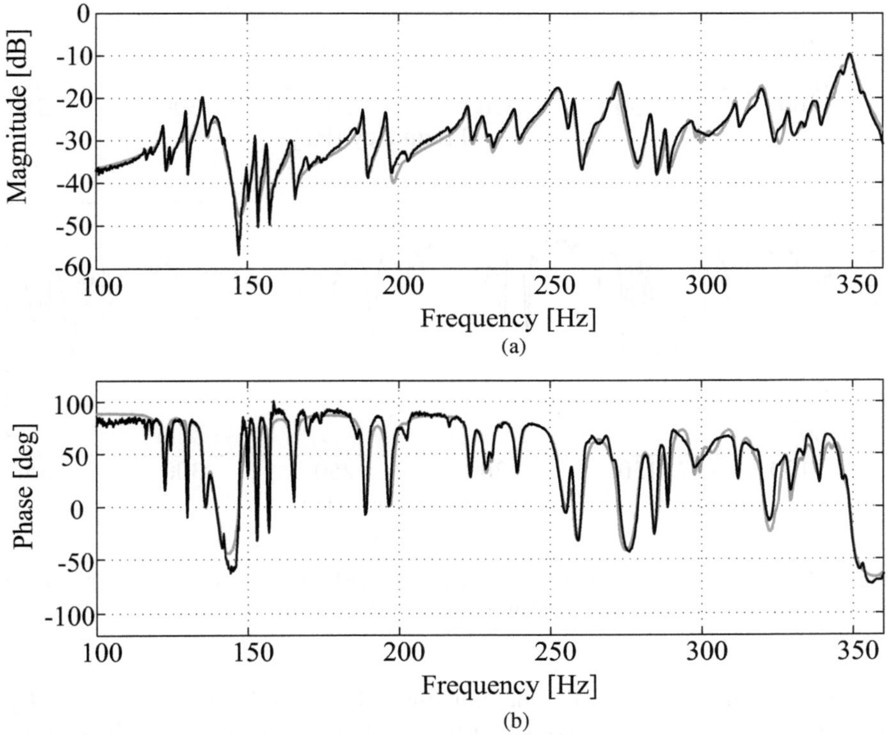

Figure 7.4 (a) Magnitude and (b) phase of the $g_{11}(j\omega)$ FRF (input voltage [V], output velocity [m/s]): measured (*solid black lines*) and identified (*solid grey lines*)

first. To this aim, the weighting matrix in Equation 3.59 was selected by taking into account the frequency range of interest up to $f_2 = 350\,\text{Hz}$. A number of trials was performed with different values of α and the corresponding reconstruction error was quantified by computing the quadratic index usually adopted for modal analysis [1], *i.e.*,

$$J_\alpha = \sum_{i \in \mathscr{Y}} \|G_i - \hat{G}(j\omega_i)\|_2^2, \tag{7.1}$$

with G and $\hat{G}$ the measured and identified FRF, respectively; the spectral lines used belong to the set $\mathscr{Y} = \{1, \ldots, 7500\} - \mathscr{I}$. The resulting index values for different values of the weight α are reported in Table 7.1.

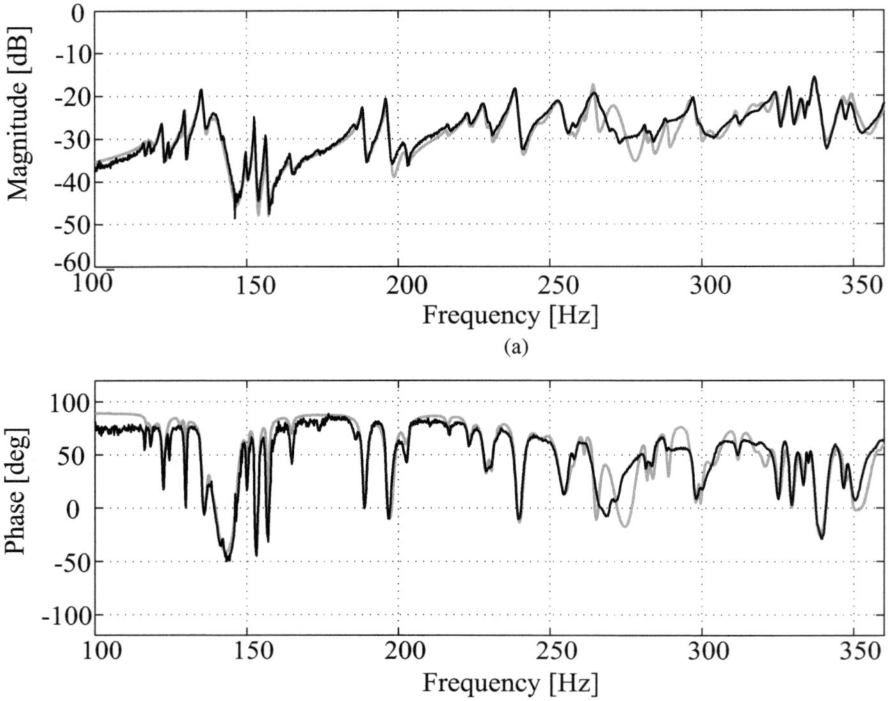

(a)

(b)

Figure 7.5 (a) Magnitude and (b) phase of the $g_{44}(j\omega)$ FRF (input voltage [V], output velocity [m/s]): measured (*solid black lines*) and identified (*solid grey lines*)

Therefore, the value $\alpha = 0.4$ that minimises the index J_α has been chosen. By applying the proposed order reduction technique, the second stage ends with the estimation of a model with only 148 controllable modes.

The identified and measured FRFs corresponding to the validation set $\mathscr{V}$ of spectral lines in the frequency range $[f_1, f_2]$, not used in the identification, are reported in Figures 7.4 and 7.5 for two colocated input–output couples and in Figures 7.6 and 7.7 for two non-colocated input–output couples. It can be recognised that the identified responses reproduce the measured ones quite well, both for magnitude and phase, despite the high modal density and the large width of the frequency range. Since it is well known that a frequency domain approach can lead to a poor result in the time domain, a validation in this domain was carried out. Figure 7.8 reports measured and modelled acceleration (computed *via* numerical differentiation of the model output velocity) for the first output in response to four different swept sine signals applied to the four input channels. The magnified portions of the plot show that the reconstruction is fairly good also in the time domain.

The algorithm proposed in Chapter 3 to tackle the identification problem for systems with non-colocated inputs and outputs has been tested for the identification of

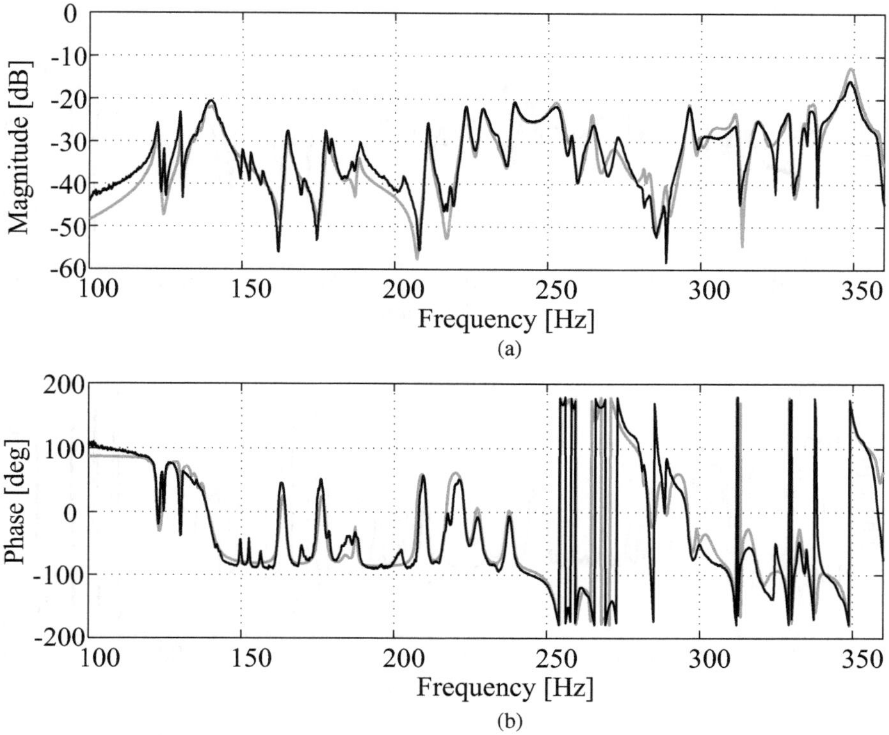

Figure 7.6 (a) Magnitude and (b) phase of the $g_{32}(j\omega)$ FRF (input voltage [V], output velocity [m/s]): measured (*solid black lines*) and identified (*solid grey lines*)

a three-output one-input system with no colocated I/O pairs. The considered system is constituted by the same aeronautical panel as above excited by a single piezoelectric patch and whose acceleration is measured in three different non-colocated points. By applying the first stage of the procedure to the identification set of spectral lines, 127 modes have been estimated, while the input and output matrices have been identified by applying the method of Section 3.4.3. The results are reported in Figure 7.9 where the measured and modelled FRFs between the input and the second output, computed on the validation set of spectral lines, are depicted. It can be recognised that the quality of the identification is quite satisfactory also when only non-colocated measurements are available.

In order to better understand the benefits of the proposed method, the identification results presented above for the 4×4 system have been compared to those achieved by four different approaches, both commercial and found in the scientific literature, *i.e.*, PolyMAX [6, 9] (hereafter named "PMAX"), FDIDENT [7] (hereafter named "FDID"), the classical subspace method in the frequency domain [8] leading to a discrete-time model (hereafter named "SSTD") and the classical subspace method in the frequency domain leading to a continuous-time model by

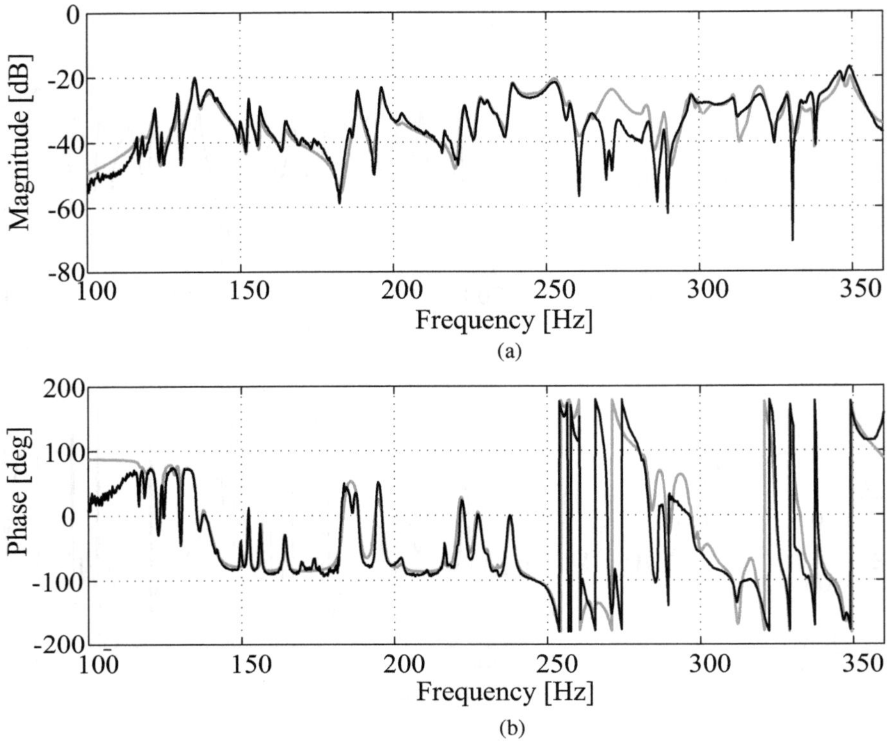

Figure 7.7 (a) Magnitude and (b) phase of the $g_{41}(j\omega)$ FRF (input voltage [V], output velocity [m/s]): measured (*solid black lines*) and identified (*solid grey lines*)

applying the bilinear transformation as proposed already in [8] (hereafter named "SSCT").

Firstly, to examine the quality of the identification of the 4×4 model of the panel a comparison has been performed, by ruling out the FDID method, since it can be applied only to SISO systems. The only way to apply the FDID algorithm to MIMO systems is to identify all the transfer functions from each input channel to each output channel separately and then build a state space realisation. On the other hand, this way, even though it provides accurate models for the single transfer functions, will certainly lead to a non-minimal realisation with such a large number of modes that it cannot be used for any practical aim. Not even a model minimisation can be easily applied, as it is well known that MIMO minimisation algorithms exhibit very poor computational characteristics for high-order systems, and this is certainly the case. In fact, the algorithm applied to each transfer function identifies different system poles even if related to the same structural mode and the algorithm does not offer any method to handle this problem, thus leading to a large number of system poles. The occurrence of this case is confirmed by the results reported in Table 7.2, where the natural frequencies estimated by FDID for two transfer functions

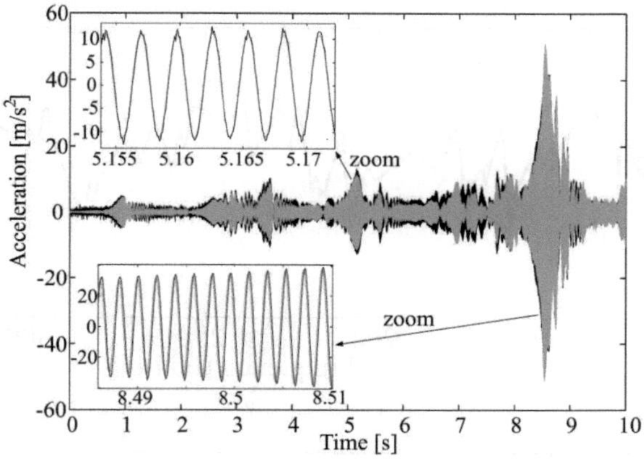

Figure 7.8 Time domain validation: measured (*solid black line*) and modelled (*solid grey line*) acceleration for first output

Table 7.2 Natural frequencies estimated by FDID in the frequency range $[200, 250]$ Hz for the transfer functions g_{11} and g_{32}

g_{11}	202.7824	216.7092	223.1541	227.8403	228.7884	232.8362	238.6911	
g_{32}	202.3531	210.6954	216.2704	223.0479	228.2799	232.9285	234.3414	238.6093

are reported in the frequency range $[200, 250]$ Hz, where several natural frequencies appear very close to each other.

The comparison among the selected MIMO methods has been carried out by computing the estimation error defined as

$$V(\omega_i) = \|G_i - \hat{G}(j\omega_i)\|_2^2, \quad i \in \mathscr{V}, \tag{7.2}$$

and the quadratic index defined as

$$J_v = \sum_{i \in \mathscr{V}} \|G_i - \hat{G}(j\omega_i)\|_2^2. \tag{7.3}$$

The estimation errors obtained with all the compared algorithms are reported in Figure 7.10 and in Table 7.3. It can be recognised that the SSDT (with a number of modes equal to the initial number set for the proposed approach, *i.e.*, $n = 165$) outperforms all the others. However, it produces a discrete-time black-box model whose parameters have no physical meaning and without the properties of minimum phase, reciprocity and a non-null D term. Moreover, if the black-box model were used to carry out standard control design procedures (*e.g.*, $\mathscr{H}_2, \mathscr{H}_\infty$ strategies), only numerical techniques could be applied to compute the controller, with all the related conditioning problems typical of high-order systems, for which closed-form solutions are difficult to obtain. In contrast, the considered model structure allows one to obtain closed-form solutions to optimal control problems as reported in Chapter 5.

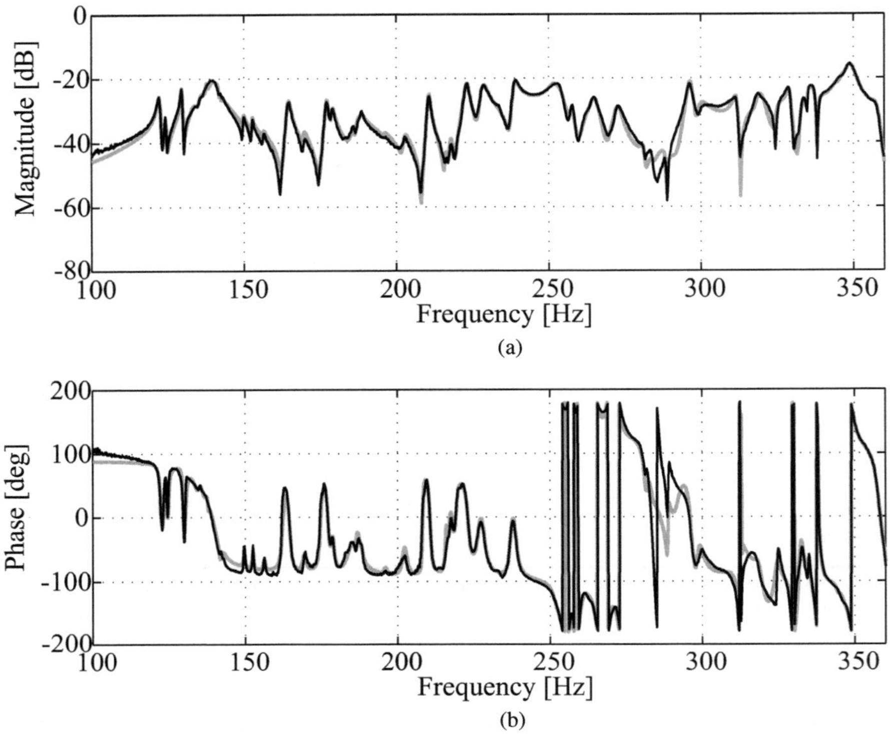

Figure 7.9 (a) Magnitude and (b) phase of the $g_{21}(j\omega)$ FRF (input voltage [V], output velocity [m/s]): measured (*solid black lines*) and identified (*solid grey lines*) in the non-colocated case

Table 7.3 Values of the quadratic index J_v in Equation 7.3 obtained with the four compared identification algorithms

algorithm	PMAX	proposed	SSDT	SSCT
J_v	15.5	9.16	1.13	35.3

If a continuous-time model is desired, the SSCT provides a model (with order equal to the initial model order set for the proposed approach, *i.e.*, $n = 165$) much less accurate than our proposed method mainly due to the inaccuracy caused by the bilinear mapping of discrete-time poles into the continuous-time. Also, the model still exhibits the same limitations as the SSDT related to the black-box approach. The quality of the identification provided by the commercial PMAX algorithm (with a number of modes equal to 120 considered optimal by the user) is the lowest among the compared techniques, but it is a grey-box approach leading to a modal model with all its structural properties. Nevertheless, it is worth noting that the PMAX method is conceived for modal analysis, which typically yields better results when used with rectangular systems (with more outputs than inputs or *vice versa*).

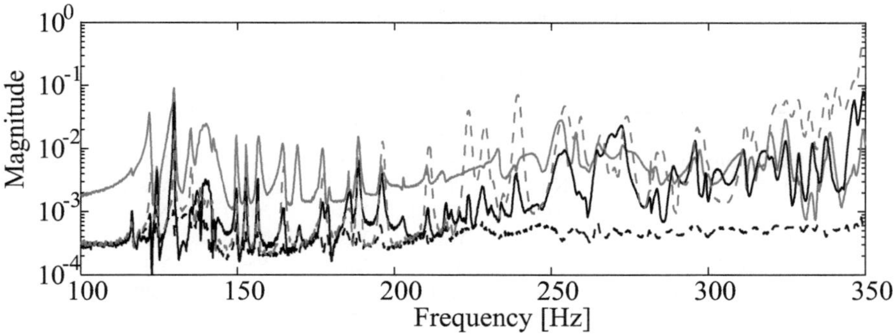

Figure 7.10 Estimation errors as defined in Equation 7.2 computed for the four compared identification algorithms: proposed (*solid black line*), SSDT (*dashed black line*), SSCT (*dashed grey line*), PMAX (*solid grey line*)

The identification methods have also been compared from a computational point of view by estimating the execution time of a typical identification session. This estimation also takes into account the interaction time required with the user, even though this highly depends on his/her expertise. The FDID algorithm, even if applied to a SISO system, requires the highest execution time, due to its iterative nature. The remaining methods are practically equivalent, but PMAX, even if computationally less demanding than the others, usually requires a long user interaction time. In fact, in the pole estimation phase, the user has to select every single mode on the so-called "stabilisation diagram" based on his/her experience, which can significantly affect the final result. Notice that the execution time of the proposed method also includes the automatic order reduction phase. The results of the comparison are summarised in Table 7.4.

7.3 Active Control of Vibrations with Piezo Actuators

A number of case studies have been carried out to assess the performance of the active vibration control algorithms presented in Chapter 5. In all the case studies discussed in this section piezoelectric ceramic actuators have been used both for producing the primary disturbance field and for generating the control forces on the structure aimed at counteracting the effects of the primary field.

In detail, the control inputs are produced using three of the four piezoelectric patches bonded on the panel surface; the output feedback acceleration is measured by means of the three colocated accelerometers placed on the rear part of the panel behind the piezoelectric actuators. The model of the structure has modes with natural frequencies in the range $\Omega = [100, 600]\,\text{Hz}$ and it was identified as described in the previous section and its parameters are reported in the Appendix. The experiments were performed by using a dSPACE rapid prototyping real-time control

Table 7.4 Comparison of the tested identification algorithms

Algorithm → / Characteristic ↓	SSDT	SSCT	FDID	PMAX[a]	proposed
MIMO	Yes	Yes	No	Yes	Yes
Guar. min. phase	No	No	Yes[b]	Yes	Yes
Grey box	No	No	No	Yes	Yes
Iterative	No	No	Yes	No[c]	No
D term	Yes	Yes	Yes[b]	No	No
Order reduction	No	No	No	Yes[d]	Yes
Execution time	med.	med.	high	med.[e]	med.

[a] The model has to be built by the user on the basis of the estimated modal parameters.
[b] The user has to impose it as a constraint.
[c] Even though the method is non-iterative, the user repeatedly uses the stabilisation diagram to assess pole estimation.
[d] Not automatic: based on the stabilisation diagram, the user selects a subset of the estimated modes.
[e] The actual time highly depends on the user expertise.

system with 16-bit A/D channels and 14-bit D/A channels at a sampling frequency of 20 kHz and described in detail in Section 6.7.

The fourth piezo is used only as a disturbance input so as to introduce a disturbance contribution not belonging to the range of the input matrix B_u in Equation 5.10. One of the known drawbacks in the application of optimal feedback control to vibration reduction problems is that the control action may focus only on the performance objectives, reducing vibrations only at the controlled points, while increasing vibration levels at other points of the structure. In order to check that the controller actually has a beneficial effect on the *whole* structure, a fourth accelerometer is used to measure the acceleration in a grid of 35 performance test points uniformly distributed on the panel (see Figure 7.11) so as to assess the global performances of the control in terms of an energy-based index defined as follows. First, the energy of the measured acceleration in the given frequency range Ω has been estimated at the kth point of the grid as

$$J_k = \sqrt{\sum_{h=1}^{M} |\Phi_k(\omega_h)|^2}, \qquad k = 1, \ldots, 35, \tag{7.4}$$

where $\Phi_k(\omega_h)$ is the acceleration spectrum measured at the kth point at the frequency $\omega_h \in \{\omega_1, \ldots, \omega_M\}$ a set of $M = 2500$ linearly spaced frequencies within Ω. Then, the relative energy reduction at the kth point has been quantified as

$$R_{k\%} = \frac{J_{k,\text{ol}} - J_{k,\text{cl}}}{J_{k,\text{ol}}} 100, \tag{7.5}$$

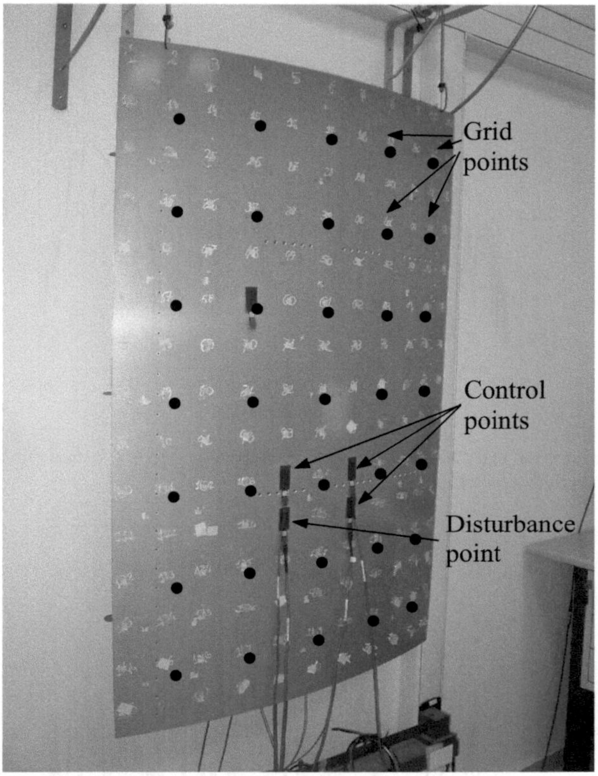

Figure 7.11 Grid of performance test points on the skin panel

where the subscripts "ol" and "cl" stand for "open loop" and "closed loop", respectively.

The experiments described in the following sections are aimed at evaluating the performance of the active vibration controllers as proposed in Sections 5.1 and 5.10. In detail, the controller based on the result of Theorem 5.2, is particularly attractive when a high modal density structure is considered, since the analytical solution to the optimal control problem is not sensitive to numerical problems typical of high-order systems. Nevertheless, this solution has a strong limit, *i.e.*, the performance weights cannot be chosen by the user to accomplish a certain objective depending on the specific application. On the other hand, the optimal controller based on the result of Theorem 5.4 allows the user to select the performance weights on the basis of the knowledge of some characteristics of the disturbance and of the structure. In order to save the desirable characteristics of the strong stability and bandpass nature of the controller, this is computed by solving an LMI problem, as explained in Section 5.10.

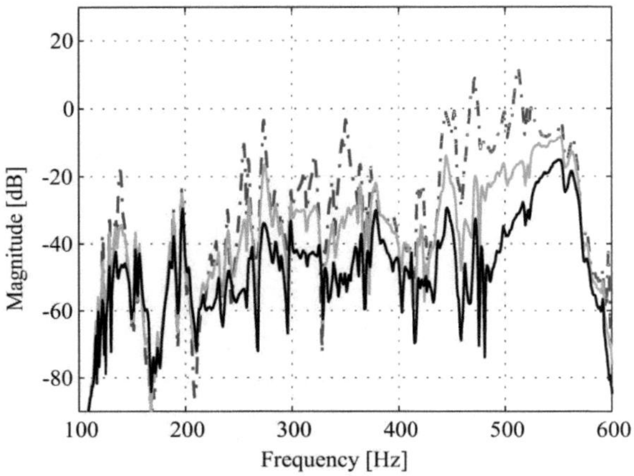

Figure 7.12 Acceleration in a colocated point of the panel: open loop (*dashed grey line*), $\mathscr{H}_2$ (*solid grey line*) and $\mathscr{H}_\infty$ (*solid black line*)

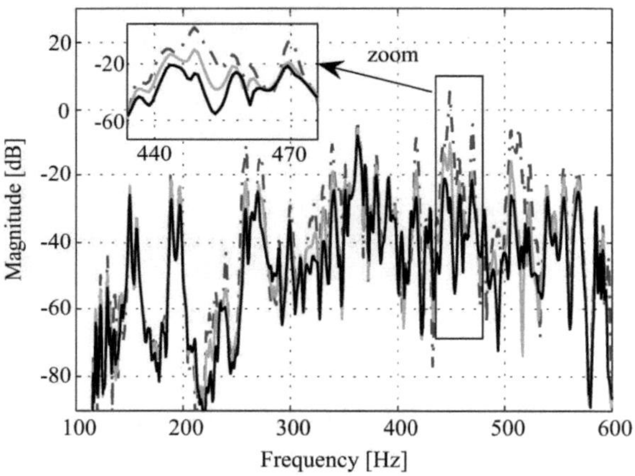

Figure 7.13 Acceleration in a non-colocated point of the panel: open loop (*dashed grey line*), $\mathscr{H}_2$ (*solid grey line*) and $\mathscr{H}_\infty$ (*solid black line*)

7.3.1 Analytically Designed Controllers

In this section the effectiveness of the $\mathscr{H}_2$ and $\mathscr{H}_\infty$ control laws reported in Section 5.1 and originally presented in [4] and [3], respectively, is shown considering their application to the selected test article.

A band-limited white noise in the range $[100, 600]$ Hz has been chosen as primary field, in order to excite all the system natural modes in the frequency range Ω. The noise has been applied to the structure on all four input channels so that the disturbance has components both in the range of B_u in Equation 5.10 and off-range. Figures 7.12 and 7.13 present results for both $\mathscr{H}_2$ and $\mathscr{H}_\infty$ case. The $\mathscr{H}_\infty$ controller

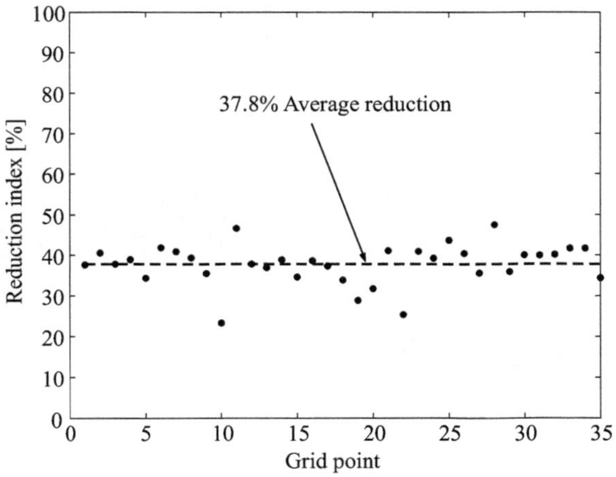

Figure 7.14 Reduction energy index at grid points of the panel with $\mathscr{H}_2$ controller

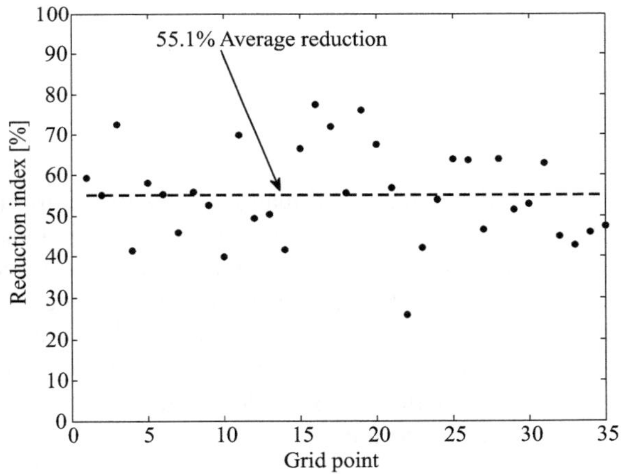

Figure 7.15 Reduction energy index at grid points of the panel with $\mathscr{H}_\infty$ controller

has been designed fixing $\gamma = 1.5$ in Equation 5.38. The figures show the acceleration spectrum measured in a control point and in a performance test point. Obviously, performances are better at the control points than at the test points, where a vibration reduction is still apparent, anyway.

Figures 7.14 and 7.15 report the values of the relative energy reduction index as defined in expression 7.5 at each grid point for the $\mathscr{H}_2$ and $\mathscr{H}_\infty$ case, respectively, showing the effectiveness of both controllers on the whole panel. The average vibration reductions are 37.8% with the $\mathscr{H}_2$ controller and 55.1% with the $\mathscr{H}_\infty$ controller almost uniformly on the whole panel, despite the small number of control actuators, the high modal density and the broadband nature of the disturbance. Figures 7.16 and 7.17 further clarify the effectiveness of the proposed controllers reporting the

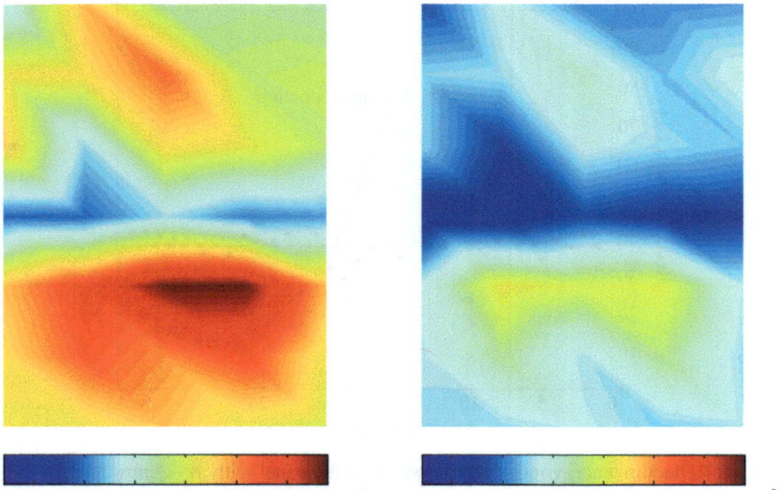

Figure 7.16 Performance of the $\mathscr{H}_2$ controller on the whole panel: open loop (a) and closed loop (b)

values of the J_k energy index defined in Equation 7.4 in all the grid points with the same colour scale in order to compare open loop with closed loop. Finally, Figure 7.18 presents the time history of a control input (piezoelectric actuator voltage) in both the $\mathscr{H}_2$ and the $\mathscr{H}_\infty$ case, showing that the control voltage with the $\mathscr{H}_\infty$ controller is generally higher than the $\mathscr{H}_2$ case. Nevertheless, in both cases the input voltage is far below the saturation limits of the actuators (200 V). This is the reason why the effects of actuator nonlinearities, as described in detail in Chapter 6, can actually be neglected in these experiments.

7.3.2 Numerically Designed Controller Using LMI

This section reports the experimental evaluation of the control law reported in Section 5.10 and applied to vibration reduction problem on the same test article as in the previous section. The same band-limited white noise of the previous section has been chosen as primary field. However, it has been applied to the structure only through the fourth piezo actuator not used as control input, thus the disturbance is totally out of the range of the input matrix B_u in Equation 5.10. This choice allows better highlighting the feature of the controller when used with additional sensors located near the entry point of the primary field and when only non-colocated feedback is used. Depending on the choice of modal performance weights (defined in

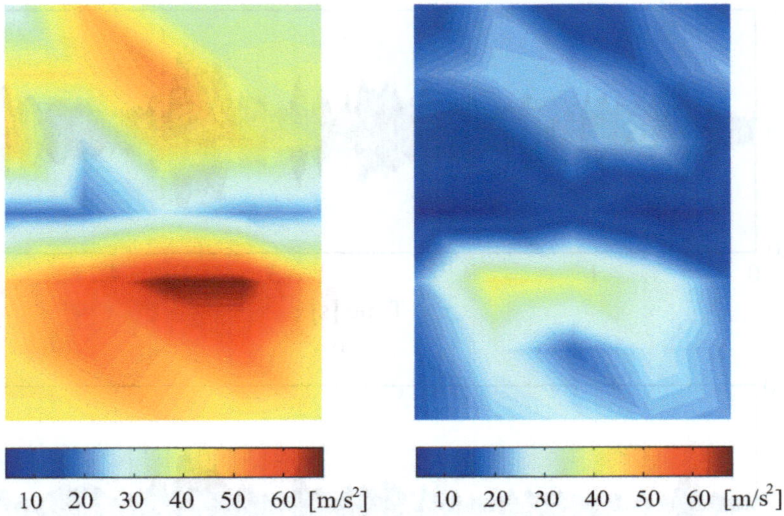

(a) (b)

Figure 7.17 Performance of the $\mathscr{H}_\infty$ controller on the whole panel: open loop (a) and closed loop (b)

Equation 5.138) and on the number of feedback sensors, the following controllers have been tested:

A. Colocated controller with uniform modal performance weights.
B. Colocated controller with modal performance weights based on the system Gramians.
C. Colocated controller with modal performance weights higher on selected modes.
D. Rectangular controller (with additional sensor) with modal performance weights based on the system Gramians.
E. Non-colocated controller with modal performance weights based on the system Gramians.

The performance obtained with the controllers above are evaluated both in terms of vibration reduction quantified according to the index in Equation 7.5 and by comparing the FRFs from disturbance input point to a control point, the disturbance point and a performance point. The following case studies summarise the performance analysis:

1. comparison between controllers A and B
2. comparison between controllers A and C
3. comparison between controllers B and D
4. evaluation of controller E

All the controllers are also compared in terms of required control energy according to the index defined below.

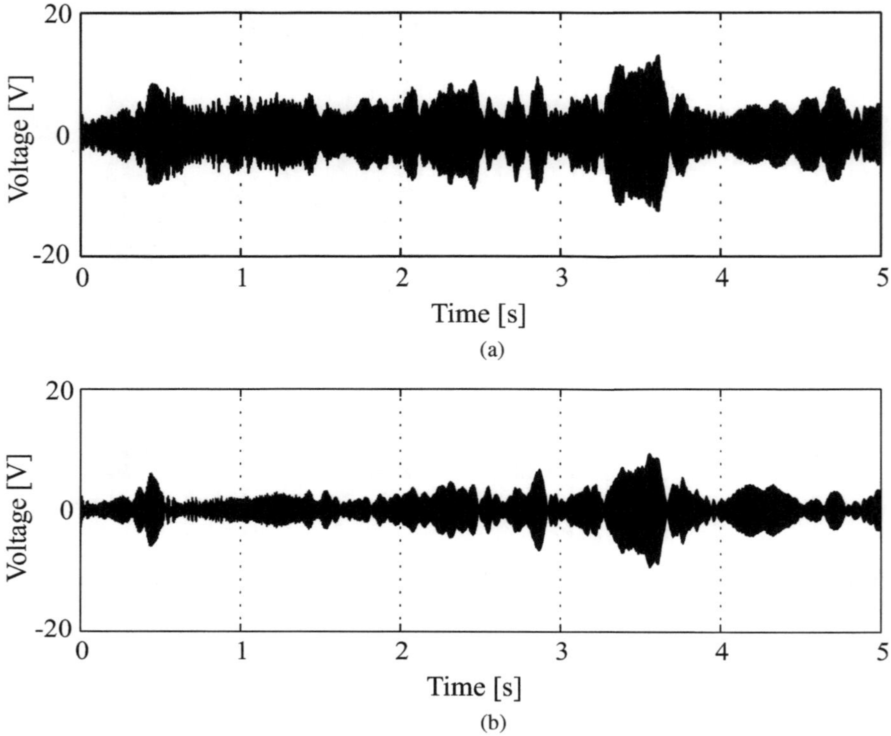

Figure 7.18 Control signals in the $\mathscr{H}_\infty$ case (a), and $\mathscr{H}_2$ case (b)

First, the energy E_s of a generic discrete-time signal $s(k)$ has been evaluated as

$$E_s = \sum_{k=1}^{N} s^2(k), \tag{7.6}$$

where N is a fixed number of samples equal for all experiments. Taking into account that the primary field is randomly generated, the energy $E^r_{u,Y}$ of the rth control signal in the experiment Y (Y denotes the control label as listed above) has been normalised with respect to the actual energy of the disturbance $E_{d,Y}$ in the same experiment, *i.e.*,

$$\hat{E}^r_{u,Y} = \frac{E^r_{u,Y}}{E_{d,Y}} \qquad r = 1,2,3, \quad Y = A,B,C,D,E. \tag{7.7}$$

The control energy required by each controller is then compared to the one required by the controller A, by considering the control energy reduction index

$$R_{e\%} = \frac{\hat{E}^r_{u,A} - \hat{E}^r_{u,X}}{\hat{E}^r_{u,A}} 100 \qquad r = 1,2,3, \quad X = B,C,D,E. \tag{7.8}$$

Table 7.5 Control energy reduction index $R_{e\%}$ for all controllers considered

	Controller B	Controller C	Controller D	Controller E
Channel 1	11.20	41.11	21.13	61.24
Channel 2	21.46	67.38	31.54	72.19
Channel 3	9.930	52.31	19.07	78.00

Table 7.6 Case study 1: values of the reduction energy index $R_{k\%}$ in Equation 7.5 obtained in the frequency range $[100, 600]$ Hz for control, disturbance and performance points

	Control pt. 1	Control pt. 2	Control pt. 3	Disturbance pt.	Performance pt.
$R_{k\%}$-controller A	74.93	77.36	67.77	29.18	26.04
$R_{k\%}$-controller B	77.49	77.88	68.25	30.74	30.58

Table 7.7 Case study 2: values of the reduction energy index $R_{k\%}$ in Equation 7.5 obtained in the frequency range $[465, 480]$ Hz for control, disturbance and performance points

	Control pt. 1	Control pt. 2	Control pt. 3	Disturbance pt.	Performance pt.
$R_{k\%}$-controller A	91.38	84.60	86.04	66.50	40.34
$R_{k\%}$-controller C	91.93	87.48	87.73	70.41	43.02

The results for all the controllers are summarised in Table 7.5, and it is evident that Controller A is the one requiring the highest energy. They will serve in the analysis of each case study carried out in the following.

Case study 1 is aimed at showing the performance of the colocated controller when no specific information is available on the disturbance. Controller A is designed according to the procedure of Theorem 5.4 by selecting uniform modal performance weights c_i in Equation 5.138, while the weighting matrices in Equations 5.113–5.114 have been fixed as $R_1 = 0.1I$, $R_{2_1} = 10^8$ and $R_{2_2} = 0.1I$. Controller B is designed by selecting the performance weights based on the system Gramians as described in Section 5.10.2 and weighting matrices $R_1 = I$, $R_{2_1} = 10^8$ and $R_{2_2} = I$. Although no information about the disturbance is used in the design of the controllers, the reduction index values reported in Table 7.6 for both controllers show a significant reduction of the vibration levels in all the points, with better performance in control points. Controller B slightly outperforms controller A in terms of vibration reduction, but with a quite lower control effort, as confirmed by the control energy reduction index values in Table 7.5. In order to verify the positive effects of the controllers in the whole frequency range, the FRFs from disturbance input to acceleration output in control, disturbance and performance points have been estimated and are reported in Figure 7.19. It is evident how the reduction is well distributed in the entire frequency range.

In case study 2, information on the frequency band of the disturbance is assumed available, *i.e.*, the disturbance spectrum can be of any type but limited to the range

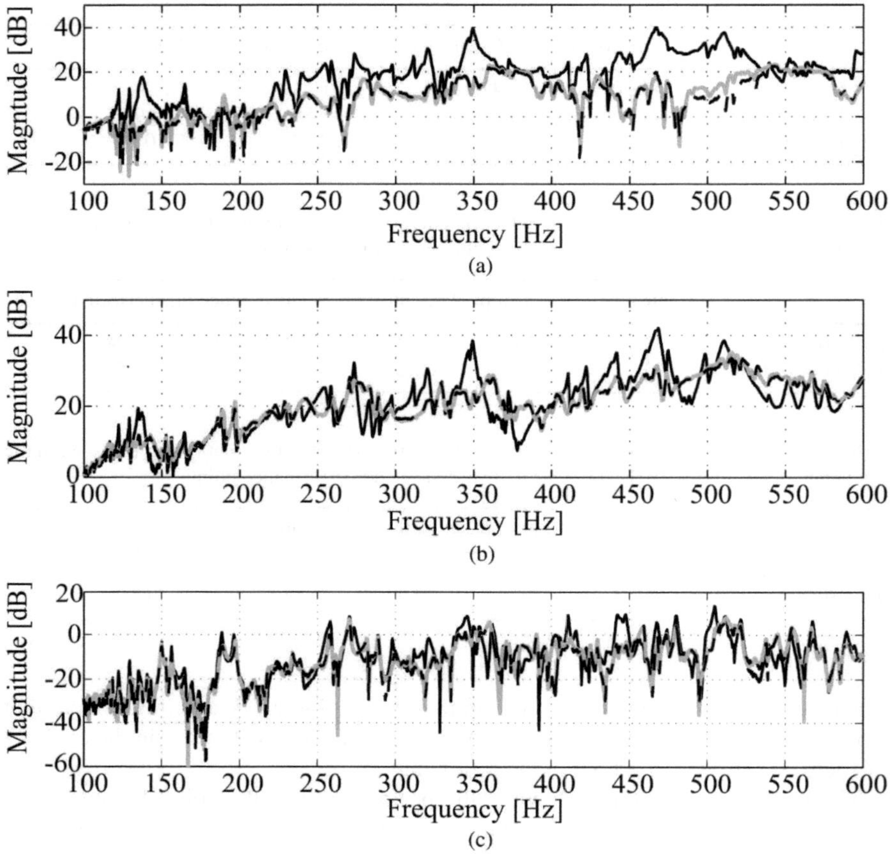

Figure 7.19 Case study 1: FRFs from disturbance input to acceleration output in a control point (a), in the disturbance point (b), in the performance point (c): open loop (*solid black lines*), controller A (*solid grey lines*), controller B (*dashed black lines*)

$\Omega_d = [465, 480]$ Hz. This information can be usefully exploited in the design phase in the selection of the modal performance weights in Equation 5.138. In order to concentrate the control effort on the modes with natural frequency within the disturbance bandwidth, to design controller C the weights for modes with natural frequency within Ω_d have been chosen higher than those used to design controller A, and uniform for all other modes. For the controller C the weighting matrices in Equations 5.113 and 5.114 have been chosen as $R_1 = 10^{-6}I$, $R_{2_1} = 10^8$ and $R_{2_2} = 10^{-6}I$. In this case, the vibration reduction has been quantified according to the index defined in Equation 7.5, where the energy of the measured acceleration has been computed only in the frequency range Ω_d. The reduction index values reported in Table 7.7 for controllers A and C show how the knowledge of the dis-

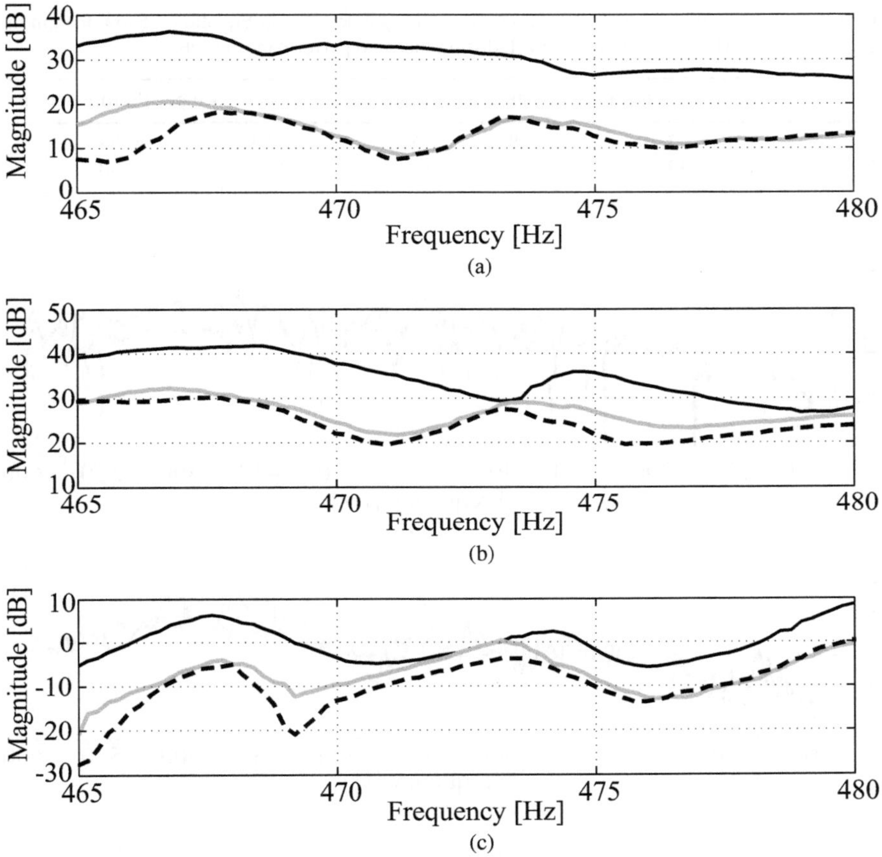

Figure 7.20 Case study 2: FRFs in the frequency range [465, 480] Hz from disturbance input to acceleration output in a control point (a), in the disturbance point (b), in the performance point (c): open loop (*solid black lines*), controller A (*solid grey lines*), controller C (*dashed black lines*)

turbance bandwidth leads to an improvement in all the control points and in the disturbance and performance points. The most relevant result obtained with controller C is the significantly lower control energy required with respect to controller A, as evident from the figures in Table 7.5. Also, the plots of the FRFs from disturbance input to acceleration output in control, disturbance and performance points in Figure 7.20 confirm the better performance of controller C with respect to controller A in the entire frequency range Ω_d.

Case study 3 tackles the problem of vibration reduction when information on the location of the application point of the disturbance force is available. Of course, in such a case, the optimal solution would be to locate an actuator–sensor pair in this entry point, which is again the situation of previous case studies. However, assume that mounting of the actuator in such a location is not possible for some

Table 7.8 Case study 3: values of the reduction energy index $R_{k\%}$ in Equation 7.5 obtained in the frequency range [100, 600] Hz for control, disturbance and performance points

	Control pt. 1	Control pt. 2	Control pt. 3	Disturbance pt.	Performance pt.
$R_{k\%}$-controller B	77.49	77.88	68.25	30.74	30.58
$R_{k\%}$-controller D	72.77	74.06	63.41	39.16	39.50

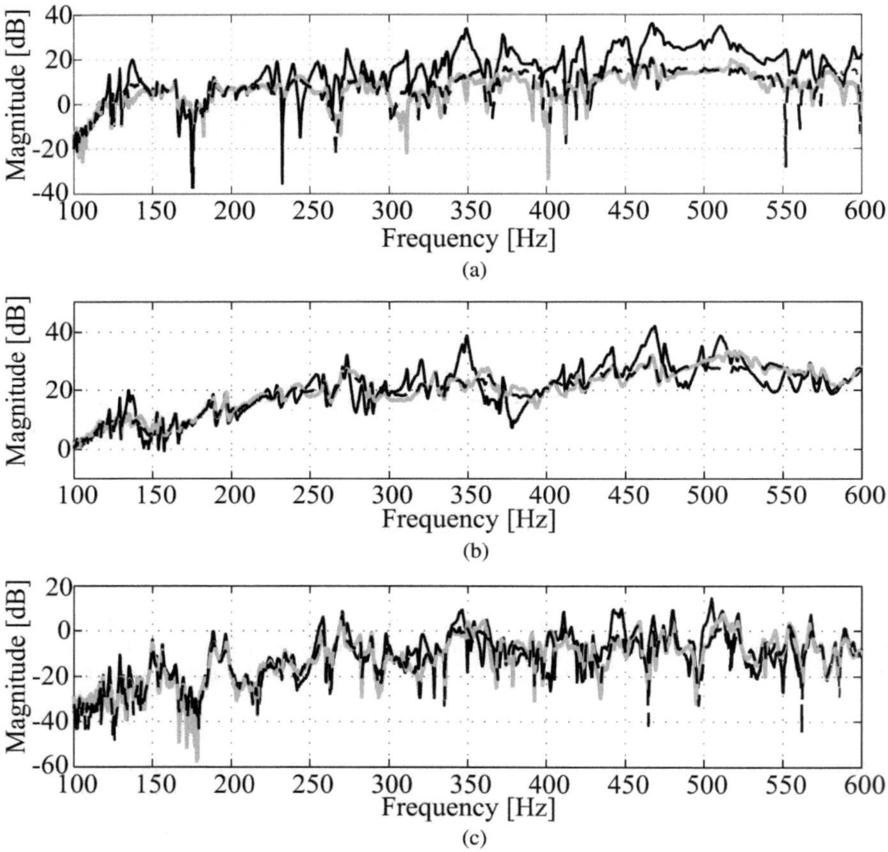

Figure 7.21 Case study 3: FRFs from disturbance input to acceleration output in a control point (a), in the disturbance point (b), in the performance point (c): open loop (*solid black lines*), controller B (*solid grey lines*), controller D (*dashed black lines*)

reason, while placement of the sensor is still allowed. In this case, the number of feedback signals is higher than the number of control points, hence the controller is rectangular. Under this assumption, controller D is designed by selecting the modal performance weights in Equation 5.138 according to system Gramians as described

Table 7.9 Case study 4: values of the reduction energy index $R_{k\%}$ in Equation 7.5 obtained in the frequency range $[100, 600]$ Hz for control, disturbance and performance points

	Control pt. 1	Control pt. 2	Control pt. 3	Disturbance pt.	Performance pt.
$R_{k\%}$-controller E	44.55	39.93	39.70	27.44	18.25

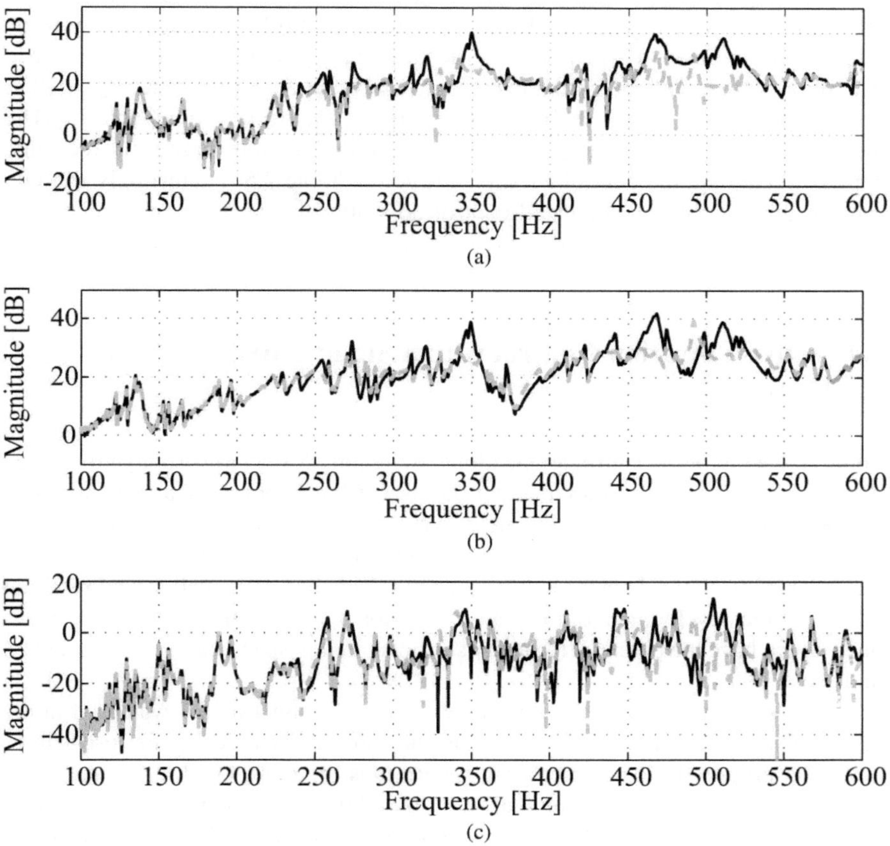

Figure 7.22 Case study 4: FRFs from disturbance input to acceleration output in a control point (a), in the disturbance point (b), in the performance point (c): open loop (*solid black lines*), controller E (*dashed grey lines*)

in Section 5.10.2 and with weighting matrices $R_1 = I$, $R_{2_1} = 1$ and $R_{2_2} = I$, and thus its performance is compared to the one of controller B, which does not exploit the additional sensor in the disturbance point. Table 7.8 shows that the reduction energy index is significantly higher in the disturbance and performance points, while a slightly degraded performance is obtained in control points. As an additional benefit, Table 7.5 shows that controller D requires less control energy than controller B.

The effects of controller D on the entire frequency range Ω is shown in Figure 7.21 where no spillover appears.

As a limit case, to assess the behaviour of the active vibration control system when the application at hand does not allow to place actuators and sensors in colocated pairs, a fourth case study was carried out. In particular, only the sensor placed in the disturbance entry point is used as feedback signal with the same three actuators of the previous case studies. The rectangular controller E of this case has no information on what happens to the structure in the control points. It is designed by selecting the weights in Equation 5.138 according to system Gramians as described in Section 5.10.2, with weighting matrices $R_1 = 10^{-4}I$, $R_{2_1} = 10^{-4}$ and $R_{2_2} = 10^{8}I$, and its performance in terms of reduction energy index is reported in Table 7.9. The reduction energy index values are lower than in the previous cases, but the same applies to the required control energy indicated in Table 7.5. Interestingly enough, even in this totally non-colocated control no spillover phenomenon occurs, as is evident from the FRFs reported in Figure 7.22.

7.4 Active Vibration Control with a Smart Auxiliary Mass Damper

Some active vibration control applications can require force levels so high that combined with weight limits they prevent the use of both conventional actuators and piezoelectric patches [5]. For example, when the actuation points are located on structural members of an aircraft fuselage, the force requirements can easily reach the order of 10 N at relatively low frequencies. This has motivated strong research efforts for the development of innovative actuators based on magnetostrictive materials, which have the potential to satisfy such demanding specifications, as explained in Chapter 6.

In this section, the controlled actuator presented in Section 6.5 was used in the active vibration control system tested on the same aeronautical panel used up to now. The proposed vibration control system must be considered advanced, since it makes use of an unconventional actuator based on a smart material with an integrated optical sensor and of hierarchical control architecture with two nested control loops. The inner loop implements the model-following approach presented in Section 6.5 in order to reduce the undesired effects of the actuator nonlinearity. The outer controller, based on the $\mathscr{H}_\infty$ algorithm of Section 5.5, is devoted to reducing the vibration caused by the disturbance. A scheme of the whole control system is shown in Figure 7.23. In this figure, P is the system to control in the grey frame, K the controller (black frame below the grey one) and the transfer function from the generalised disturbance w to the generalised output z is given by the lower linear fractional transformation $T_{zw} = \mathscr{F}_l(P,K)$. The vibration controller was designed on the basis of the optimal control strategy presented in Section 5.1.

The real aeronautical flexible structure used to test the proposed strategy is the fuselage skin panel shown in Section 7.1. The actuator was mounted to a stiffener by

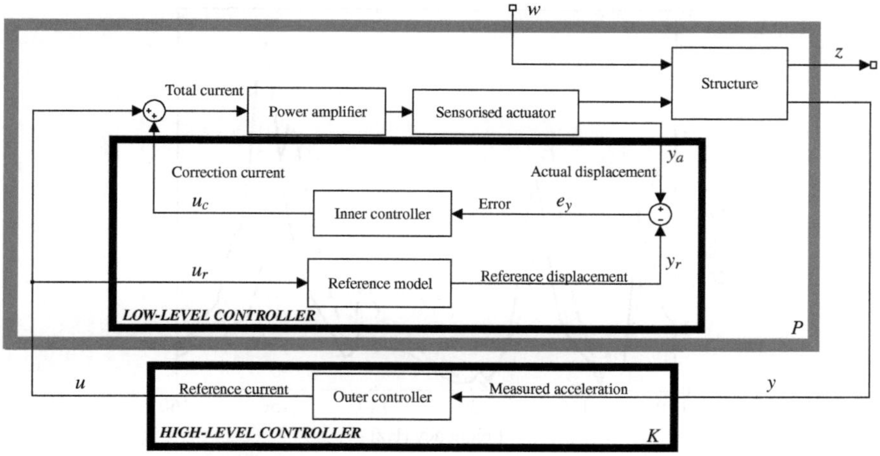

Figure 7.23 Advanced control scheme for vibration control

Figure 7.24 Smart actuator mounted on the structure

removing a rivet (see Figure 7.24). The output acceleration was measured by means of an accelerometer placed on the rear part of the panel in correspondence with the actuator mounting point (colocated case).

The most relevant effect of actuator nonlinear kinematics when it is mounted on the vibrating structure is a dependence of the natural frequencies of the structure on the amplitude of the input current. This effect is particularly relevant in the frequency range close to the resonant frequency of the actuator. Figure 7.25 reports the frequency response function (FRF) measured from the input current to the optically sensed displacement using different amplitudes of the exciting chirp signal, *i.e.*, 0.6, 1.2, 2 A. The shift of the natural frequencies appears evident.

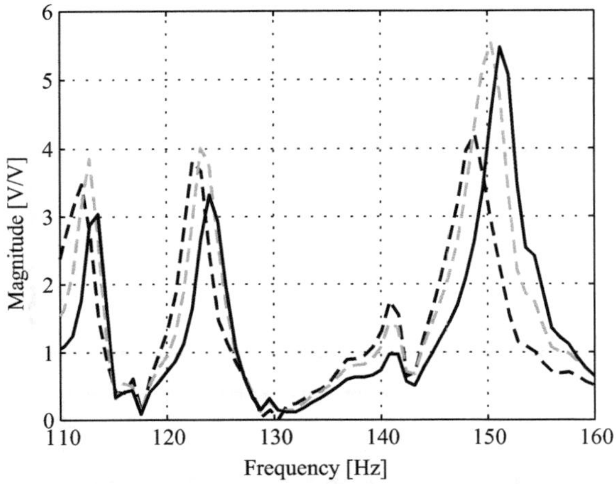

Figure 7.25 Structure FRF measured at different actuator current amplitude: 0.6 A (*solid black line*), 1.2 A (*dashed grey line*), 2 A (*dashed black line*)

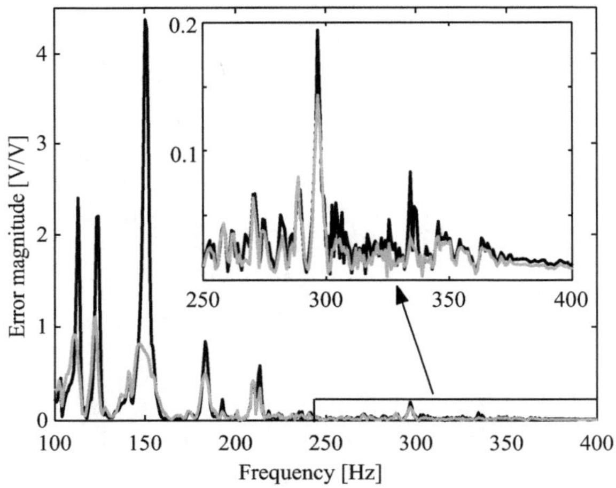

Figure 7.26 Indicator $E(\omega)$ as defined in Equation 7.9, with actuator controller switched off (*solid black line*) and on (*solid grey line*)

As first step of the design procedure, the linear model in Equations 6.16 and 6.17 has been identified by resorting to the FDIDENT Matlab Toolbox, resulting in a 24-order system. Using this system as reference model, the LQG controller in Equations 6.22 and 6.23 has been designed by experimentally tuning the weighting matrices of the Riccati equations. In order to verify the effect of the designed controller, the following indicator has been defined to compare the FRFs estimated at two different levels of the input current

$$E(\omega) = |H_{0.6}(j\omega) - H_{1.2}(j\omega)|, \tag{7.9}$$

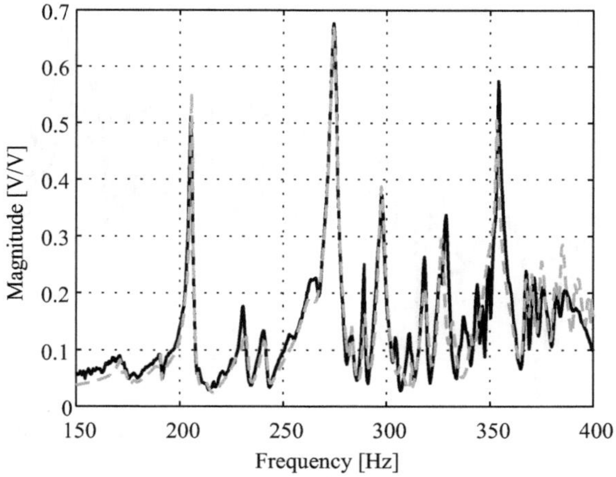

(a)

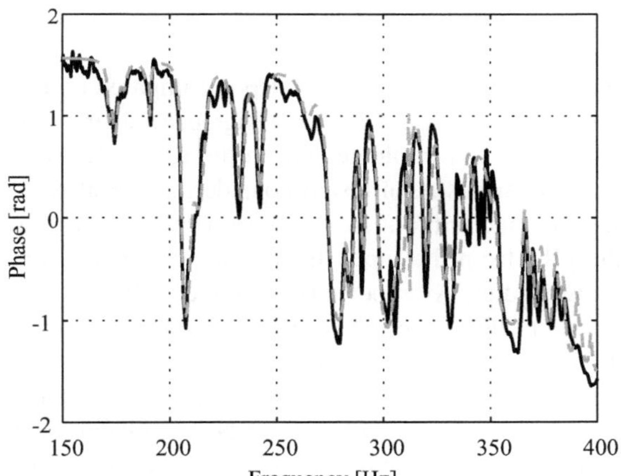

(b)

Figure 7.27 Experimental identification ((a) magnitude and (b) phase) of the structure: measured FRF (*solid black lines*) and modelled FRF (*dashed grey lines*)

where $H_{0.6}(j\omega)$ and $H_{1.2}(j\omega)$ are the FRFs measured with an amplitude of the input chirp current signal equal to 0.6 and 1.2 A, respectively. The results are shown in Figure 7.26, which shows the indicator E evaluated with the actuator controller switched off and on. For the whole frequency range the indicator assumes lower values when the actuator controller is activated, meaning that the alteration of the structural response due to the nonlinearity of the actuator is significantly reduced.

The second step of the design procedure is devoted to identifying a model of the plant P necessary for computing the $\mathscr{H}_\infty$ vibration controller presented in Section 5.1. By applying the grey-box identification procedure of Section 3.4 a model

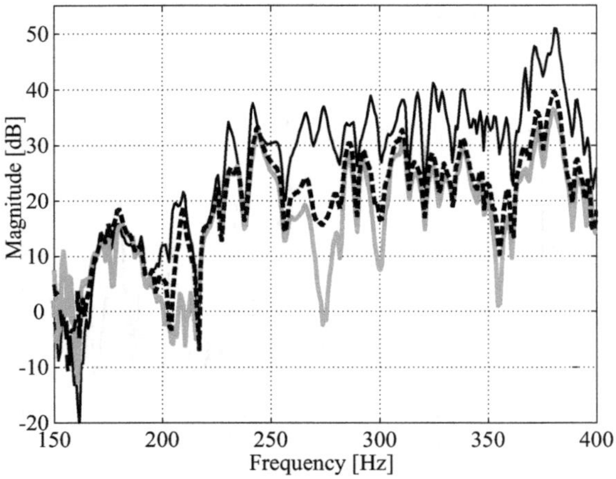

Figure 7.28 Experimental results for the two-level control scheme: inner OFF/outer OFF (*solid black line*), inner OFF/outer ON (*dashed black line*), inner ON/outer ON (*solid grey line*)

with 35 modes has been estimated, whose FRF is compared to the measured FRF in Figure 7.27. Choosing the value 1.5 for the design parameter γ in Equation 5.38, the resulting controller has been implemented on the dSPACE system. The complete advanced control system has been tested considering the effects of the high-level (outer) controller with and without the low-level (inner) control. In Figure 7.28 the magnitude of the FRF from the input disturbance to the measured acceleration is reported in the case studies. As expected, the higher values correspond to the open loop (inner OFF/outer OFF) case, while the better disturbance rejection (vibration reduction) is obtained when both feedback controllers are active (inner ON/outer ON), with a reduction of 14.4dB of the infinity norm. Comparing the case inner ON/outer ON and the case inner OFF/outer ON it is evident as the performances of the whole system can significantly improve, exploiting the full potential of the smart actuator.

7.5 Conclusions

This chapter aimed at demonstrating effectiveness of the identification and control techniques proposed by the authors in Chapters 3 and 5, respectively. The test article was selected as a real aeronautical stiffened panel, with high modal density and subject to broadband disturbances in order to test both the identification and control strategies in a case study very close to a real-world application. Initially, the chapter showed the capabilities of the identification procedure, which was compared to other existing methods. Then, the identified model was used for the controller design tested in the second part of the chapter. A complete set of control experiments was carried out, ranging from colocated to non-colocated control. The last section presented experimental results for active vibration reduction using an innovative magnetostrictive actuator.

References

1. Balmès E (1996) Frequency Domain Identification of Structural Dynamics Using the Pole/Residue Parametrization. In: Proc Int Modal Analysis Conf, Dearborn, MI
2. Cavallo A, De Maria G, Natale C, Pirozzi S (2007) Gray-box Identification of Continuous-time Models of Flexible Structures. IEEE Trans Contr Syst Tech, 15:967–981
3. Cavallo A, De Maria G, Natale C, Pirozzi S (2008) Robust Control of Flexible Structures with Stable Bandpass Controllers, Automatica, 44:1251–1560
4. Cavallo A, De Maria G, Natale C, Pirozzi S (2006) Model Identification and Optimal $\mathscr{H}_2$ Vibration Control of an Aeronautical Panel, In: Proc The Sixth International Symposium on Active Noise and Vibration Control, Adelaide.
5. Cavallo A, May C, Minardo A, Natale C, Pagliarulo P, Pirozzi S (2009) Active Vibration Control by a Smart Auxiliary Mass Damper Equipped with a Fiber Bragg Grating Sensor. Sens and Act A: Physical 153:180–186
6. Guillaume P, Verboven P, Vanlanduit S, van der Auweraer H, Peeters B (2003) A Poly-reference Implementation of the Least-squares Complex Frequency-domain Estimator. In: Proc Int Modal Analysis Conf, Kissimmee (FL)
7. Kollár I (1993) On Frequency-Domain Identification of Linear Systems. IEEE Trans Instrum and Meas 42(1):2–6
8. McKelvey T, Akçay H, Ljung L (1996) Subspace-Based Multivariable System Identification from Frequency Response Data. IEEE Trans Automatic Control, 41:960–979
9. Peeters B, Guillaume P, Van der Auweraer H, Cauberghe B, Verboven P, Leuridan J (2004) Automotive and Aerospace Applications of the PolyMAX Modal Parameter Estimation Method. In: Proc Int Modal Analysis Conf, Dearborn, MI
10. Welch PD (1967) The Use of Fast Fourier Transform for the Estimation of Power Spectra: A Method Based in Time Averaging Over Shot, Modified Periodograms. IEEE Trans Audio Electroacoustics AU-15:70–73

Chapter 8
Conclusions and Future Developments

8.1 Conclusions

The book presented a complete road map, from modelling to implementation, towards the objective of vibration reduction in flexible structures through active control techniques. This problem, as recalled in the Introduction of the book, is of interest in many application fields. Despite the different nature of the structure and the source of the vibration source, as clarified in the chapter devoted to modelling, the active control problem can be addressed by resorting to a common methodology, *i.e.*, the model-based approach.

Chapter 2 showed how any flexible structure, independently of its nature, dimension and geometry, can be mathematically modelled in a unique way through the formulation and solution of a general boundary value problem. To allow the control engineer to design a model-based controller to implement in a real control system, the resulting infinite dimensional model was approximated according to one of the available techniques. The structural properties of the finite-dimensional model were highlighted with the aim of exploiting them in the following phases of the controller design. The first phase was the experimental identification of the dynamic model, performed for a twofold reason. Firstly, alternative modelling techniques like FEM usually lead to a model unsuitable for control design purpose due to its large dimensionality. Secondly, an experimental identification procedure directly provides a parametrised model describing the behaviour of the real system, whereas a numerical FE model must always be experimentally correlated, thus requiring additional effort.

The identification procedure proposed in Chapter 3 is based on a grey-box approach leading to a model whose parameters have a clear physical meaning with the same properties of a flexible structure model, and a complexity compatible with the successive control design phase. The procedure was presented in an algorithmic fashion so as to allow the interested reader to easily implement it.

Chapters 4 and 5 offered a wide discussion concerning different control strategies proposed to address the vibration reduction of flexible structures. Two shapes of signals were considered, namely tonal and broadband disturbances. Concerning the structure, the distinction between systems with low and high modal densities was outlined. The combination of the different disturbances-structure led to a discussion of four main problems. In Chapter 4, internal model strategies, possibly with adaptive capabilities, were considered for systems with tonal disturbances and low modal densities. Sliding manifold approaches with robust observers were considered for systems with low modal densities, both for tonal and broadband disturbances. In Chapter 5, optimal approaches with $\mathscr{H}_2$ and $\mathscr{H}_\infty$ cost indices were considered for high modal density structures, even in the most complex broadband disturbance case.

The book dedicated a special focus to the implementation issues in Chapter 6. A review of the most commonly used smart materials for actuation purposes was carried out, with an in-depth examination of piezoelectric ceramics and magnetostrictive materials. The characteristics of each smart actuator were critically discussed from the control engineering point of view. The hysteresis' nonlinearity typical of such smart materials was suitably modelled and a technique for its compensation was presented. An innovative magnetostrictive auxiliary mass damper was also described and a special version with an embedded optical sensor was tested with a low-level feedback controller so as to reduce its nonlinearity, thus making it usable in a higher level feedback control loop directly devoted to vibration reduction. The last section of the chapter was devoted to describing the architecture of the rapid control prototyping system used to experimentally test the active control strategies.

The results of the tests were described in detailed in Chapter 7, where an aeronautical panel was selected as the test article and both piezoelectric and magnetostrictive actuators were tested. The quality of the model obtained by using the grey-box identification procedure was compared to that of models identified by using techniques well known in the literature or implemented in commercial software packages. The effectiveness of the active control algorithms was assessed in terms of both performance and required control effort.

The conclusions that can be drawn from the work done and the results obtained are summarised as follows. The application of the active control strategies proposed in this book to a real vibration reduction problem can be considered successful. The reason for this success can be ascribed to the many positive aspects of the resulting control system. First, the controller, designed on the basis of a good quality dynamic model of the flexible structure, can been directly obtained in an analytic form, or the numerical problem to be solved (*i.e.*, the LMI problem in Theorem 5.5) is well conditioned and its complexity scales up only quadratically with the system dimensionality. Second, every controller guarantees strong stabilisation of the system, thus enhancing robustness and simplifying implementation. Furthermore, the controller can easily be implemented with any number of sensors located anywhere. This is a very important characteristic in many real world problems, when location of a sensor-actuator pair is infeasible or the adoption of additional sensors can provide better understanding of the vibrational state of the structure. Finally, the latest findings on the selection of design weights, based on system Gramians or on individual modes selections, have been demonstrated to better distribute the control effort in the frequency range of interest and can pave the way to future developments in the field of active noise control, as clarified in the next section.

Of course, as any engineering result, the control approach summarised above has a number of limitations that require further developments to be overcome. The first limitation is due to the assumption of a linear behaviour of the structure, and this limits the admissible magnitude of the structure vibrations. Fortunately, in most of the applications this assumption is not very conservative. On the other hand, in an active control system, there are other sources of nonlinearities, mostly coming from the actuation system, especially when smart materials are adopted. Nevertheless, the book proposed some countermeasures that can be adopted, *e.g.*, hysteresis compensation or low-level feedback control loops. The second limitation is the model-based approach, which means that the design can be carried out only if a model is available, therefore the application of an identification procedure is mandatory. Also, the dimension of the model grows linearly with the modal density of the flexible structure and with the frequency bandwidth of interest. However, the most limiting factor of the active vibration control approach is certainly related to the complexity of the whole control system. In fact, it comprises at least the actuators, the sensors (in number equal to or greater than the number of actuators) the power amplifiers and the control hardware, which includes a computational unit and I/O boards. This high number of components has three major drawbacks, *i.e.*, the limited reliability, which usually decreases as the number of components increases, the high costs and the possibly large weight of the overall system. If the last issue is certainly the main limiting factor in aeronautic applications, it could be of less importance in different application fields, *e.g.*, ground transportation systems or civil engineering.

8.2 Future Developments

As has just been recalled, active vibration control systems have a strong potential impact also on application fields different from the aeronautical industry; the entire modelling, design and implementation process proposed in this book could be successfully adopted to address vibration reduction problems in such sectors. For example, the problem of controlling civil structures subjected to earthquakes could be tackled, of course by a suitable selection of the actuation system. The same applies for the improvement of riding comfort of high-speed trains, where the reduction of the vibration levels is of great interest. Concerning the improvement of passenger comfort in transportation systems, another challenge is the noise reduction inside the vehicle. To this aim, the identification and control techniques proposed in the book could be usefully exploited towards this objective. In fact, a well known noise reduction strategies is the *active structural acoustic control*, which is based on the idea to act on the source of the noise, *i.e.*, the flexible structure. This approach has been demonstrated to be successful in a real noise control problem, like those tackled in the MESA and MESEMA project or other EU funded research projects. Therefore, by extending the identification procedure to the case of modelling the coupled acoustic–structural dynamics, it could be possible to obtain a model useful for control design. The design guidelines that can be forecast could rely on the possibility to control individual modes offered by the specific control algorithm presented in Chapter 5. Moreover, both for vibration control and noise control problems, a critical issue is the selection of the actuators and sensors locations. Exploiting the identification procedure applied for experimental modal analysis as presented in Chapter 3, it could allow one to tackle the optimal placement of actuators and sensors, by minimising a cost function depending on the specific application. The same analysis could be also used to detect the key points on the structure where to place passive or semi-active devices. Finally, owing to the unsupervised nature of the identification procedure, it could be the starting point of a real-time health monitoring system.

Appendix
Dynamic Model of the Aeronautical Panel

This appendix is intended to provide the dynamic model of the aeronautical panel chosen in the Chapter 7 as the test article to readers interested in reproducing the results presented throughout the book, or to test their own control algorithms. The model is written in modal coordinates, with 148 modes, and has been identified according to the procedure reported in Chapter 3 whose results are also described in detail in Chapter 7. The model equations are the following:

$$\begin{aligned}
\dot{x}(t) &= \begin{pmatrix} 0 & I \\ -\Omega & -\Lambda \end{pmatrix} x(t) + \begin{pmatrix} 0 \\ B_{u2} \end{pmatrix} u(t) + \begin{pmatrix} 0 \\ B_{d2} \end{pmatrix} d(t), \\
y_f(t) &= \begin{pmatrix} 0 & B_{u2}^{\mathrm{T}} \end{pmatrix} x(t), \\
y_d(t) &= \begin{pmatrix} 0 & B_{d2}^{\mathrm{T}} \end{pmatrix} x(t),
\end{aligned}$$

where the inputs and outputs have the following meaning with the measurement units indicated in Figure A.1, where the block scheme of the overall control system is depicted:

- $u(t) \in \mathbb{R}^3$ is the vector of voltages at the input of the power amplifiers driving the three piezo actuators in the control points;
- $d(t) \in \mathbb{R}$ is the voltage at the input of the power amplifier driving the piezo actuator in the disturbance point;
- $y_f(t) \in \mathbb{R}^3$ and $y_d(t) \in \mathbb{R}$ are the velocities in the control and disturbance points, respectively.

The diagonal matrices $\Lambda \in \mathbb{R}^{148\times 148}$ and $\Omega \in \mathbb{R}^{148\times 148}$ have the following structure:

$$\begin{aligned}
\Lambda &= \mathrm{diag}\{2\zeta_1\omega_{n_1}, \ldots, 2\zeta_{148}\omega_{n_{148}}\}, \\
\Omega &= \mathrm{diag}\{\omega_{n_1}^2, \ldots, \omega_{n_{148}}^2\},
\end{aligned}$$

with $\zeta_l \in (0,1)$ and ω_{n_l} the damping coefficient and the natural frequency of the $l-$th mode, respectively, whose values are reported in Table A.1. The control input

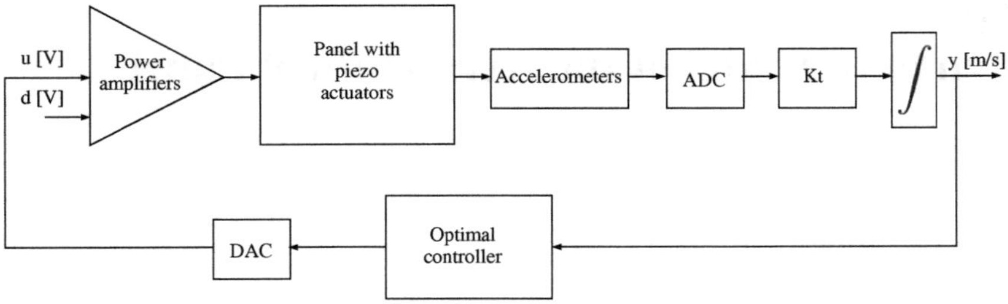

Figure A.1 Block scheme of the active vibration control system

matrix $B_{u2} \in \mathbb{R}^{148\times 3}$ entries are reported in Table A.2, while the entries of the output disturbance matrix $B_{d2}^{\mathrm{T}} \in \mathbb{R}^{1\times 148}$ are reported in Table A.3[1].

In Figure A.1 the inputs and outputs of the model are obvious, where y denotes the vector obtained by stacking the outputs y_f and y_d. Note that the output velocities are directly available in m/s owing to the cascade of the transduction constant

$$K_t = (0.1 \cdot 100 \cdot 10^{-3})^{-1}$$

and the integrator. The transduction constant converts the normalised voltage output by the ADC converter in m/s^2, by taking into account the accelerometer sensitivity in Table A.6 and the normalising factor 0.1 of the ADC converter. The integrator is numerically implemented by converting, *via* the Tustin method, at the sampling frequency of the whole control loop 25 kHz, the following transfer function in the Laplace domain

$$I(s) = \frac{\tau_1}{1+s\tau_1}\frac{\tau_2^3 s^3}{(1+s\tau_2)^3},$$

where the time constants have the values $\tau_1 = 4.8\,\mathrm{s}$ and $\tau_2 = 10^4\tau_1$. This choice allows one to accurately estimate the time integral of the acceleration in the frequency range of interest. All the technical characteristics of the control system components, *i.e.*, piezo actuators, power amplifier, accelerometers, are reported in Tables A.4–A.6. The details of the rapid prototyping real-time control system can be found in Chapter 6.

[1] Of course, the reader interested in using this model to test the algorithms presented in the book or his own algorithm can obtain the data in electronic format by contacting the authors.

Table A.1 Modal parameters ζ and ω_n for all 148 modes

Mode	$\zeta[10^{-2}]$	ω_n[rad/s]	Mode	$\zeta[10^{-2}]$	ω_n[rad/s]	Mode	$\zeta[10^{-2}]$	ω_n[rad/s]
1	0.16	733.480	**51**	0.64	2020.60	**101**	0.26	2773.47
2	0.50	746.880	**52**	0.31	2040.43	**102**	0.67	2779.14
3	0.36	768.800	**53**	0.24	2066.99	**103**	0.25	2794.15
4	0.21	781.960	**54**	1.91	2083.10	**104**	0.19	2805.78
5	0.21	815.160	**55**	0.30	2091.42	**105**	0.28	2827.23
6	0.29	848.080	**56**	0.48	2093.30	**106**	0.38	2831.88
7	0.48	850.400	**57**	0.11	2105.13	**107**	0.19	2844.32
8	0.91	869.080	**58**	0.28	2119.52	**108**	0.50	2853.92
9	1.66	878.210	**59**	0.29	2128.19	**109**	0.34	2873.90
10	0.20	941.440	**60**	0.19	2170.96	**110**	0.28	2891.58
11	0.17	959.390	**61**	0.33	2176.22	**111**	0.21	2915.27
12	0.23	983.230	**62**	0.84	2182.97	**112**	0.22	2935.88
13	0.39	1034.43	**63**	0.45	2192.19	**113**	0.91	2937.69
14	0.41	1065.47	**64**	0.24	2193.18	**114**	0.20	2941.63
15	0.30	1113.28	**65**	0.78	2199.67	**115**	0.49	2944.38
16	0.30	1124.19	**66**	0.24	2217.59	**116**	0.29	2945.86
17	0.61	1167.56	**67**	0.38	2248.06	**117**	0.21	2977.56
18	0.54	1179.18	**68**	1.15	2269.26	**118**	0.26	3010.58
19	0.25	1183.50	**69**	0.25	2273.00	**119**	0.24	3042.57
20	0.31	1231.70	**70**	0.56	2282.92	**120**	0.26	3063.36
21	0.45	1273.30	**71**	0.26	2285.71	**121**	0.50	3090.74
22	0.21	1323.66	**72**	0.63	2301.99	**122**	0.32	3127.23
23	0.34	1361.26	**73**	0.14	2314.05	**123**	0.81	3152.00
24	0.29	1366.75	**74**	0.35	2322.02	**124**	0.17	3163.38
25	0.34	1401.69	**75**	0.14	2335.44	**125**	0.64	3174.75
26	0.49	1434.23	**76**	0.21	2341.00	**126**	0.20	3203.52
27	1.30	1443.24	**77**	0.36	2359.22	**127**	0.58	3209.61
28	0.42	1449.08	**78**	0.21	2364.11	**128**	0.13	3214.83
29	0.24	1468.66	**79**	0.51	2365.57	**129**	0.27	3245.25
30	0.38	1499.59	**80**	0.39	2368.97	**130**	0.28	3255.45
31	0.76	1591.30	**81**	0.32	2401.28	**131**	0.15	3273.52
32	0.29	1619.60	**82**	0.25	2440.31	**132**	0.63	3298.17
33	0.61	1651.83	**83**	0.13	2454.18	**133**	0.15	3327.92
34	0.57	1663.69	**84**	0.68	2456.90	**134**	0.46	3361.21
35	0.99	1687.57	**85**	0.36	2476.84	**135**	0.26	3380.67
36	0.38	1696.42	**86**	0.16	2487.63	**136**	0.55	3399.58
37	1.79	1707.86	**87**	0.44	2492.14	**137**	0.21	3465.07
38	0.50	1713.95	**88**	0.27	2520.16	**138**	0.37	3485.21
39	0.23	1766.56	**89**	0.55	2535.81	**139**	0.64	3505.11
40	0.30	1781.41	**90**	0.16	2577.75	**140**	0.16	3531.71
41	0.19	1812.31	**91**	0.28	2617.51	**141**	0.17	3551.33
42	0.82	1814.55	**92**	0.18	2632.24	**142**	0.59	3557.17
43	0.31	1860.75	**93**	0.86	2635.17	**143**	0.17	3562.44
44	1.83	1862.44	**94**	0.25	2651.78	**144**	0.53	3602.26
45	0.33	1868.72	**95**	0.18	2655.86	**145**	0.35	3653.43
46	0.57	1880.64	**96**	0.78	2694.95	**146**	0.25	3675.12
47	0.25	1958.32	**97**	0.16	2728.29	**147**	0.14	3699.24
48	1.13	1983.33	**98**	0.29	2739.29	**148**	0.23	3730.45
49	0.47	1996.93	**99**	0.38	2756.03			
50	0.47	2012.34	**100**	0.52	2770.95			

Table A.2 Input submatrix $B_{2u} \in \mathbb{R}^{148\times 3}$

Rows from 1 to 50			Rows from 51 to 100			Rows from 101 to 148		
0.072	0.000	0.000	−0.561	−0.449	0.907	1.608	0.165	−0.341
0.070	0.000	0.000	−0.000	−0.000	1.202	0.000	−0.670	0.779
−0.415	0.447	0.475	−0.609	−0.646	0.906	−0.765	0.720	0.582
−0.146	0.210	0.149	1.120	0.000	1.105	0.361	−0.523	0.930
0.427	−0.412	−0.403	0.103	0.336	1.504	−0.000	−1.407	1.667
0.106	−0.000	0.072	−0.084	0.321	0.000	0.969	0.000	2.231
0.765	−0.313	−0.098	0.000	0.333	0.642	0.000	−0.323	0.889
−0.560	0.803	0.837	−0.606	−0.866	1.193	0.420	−0.341	0.678
−1.052	1.168	1.354	0.654	0.191	0.544	−0.000	0.334	0.000
0.013	0.000	−0.080	0.616	0.555	−0.255	0.974	−0.000	−0.000
0.282	0.000	0.000	0.965	−0.000	−0.400	0.000	0.240	0.416
0.287	0.000	0.000	0.939	1.590	0.000	−1.284	0.958	−0.000
0.378	0.622	0.423	1.684	2.117	−1.548	−3.047	3.988	−0.727
−0.000	0.130	−0.000	1.163	1.033	−0.776	1.105	1.309	−0.914
−0.000	0.476	0.479	−1.074	0.000	0.476	1.250	0.000	2.794
0.000	0.333	0.316	0.344	0.000	0.193	−0.343	1.313	−0.000
0.347	−0.294	−0.242	−0.000	0.737	0.821	1.188	−0.000	0.000
0.000	0.344	0.180	−0.116	0.000	0.000	0.707	−0.000	0.000
0.599	−0.153	−0.238	−0.322	0.000	0.935	−0.343	0.000	−0.000
0.628	−0.000	0.000	−0.000	−0.600	1.548	0.362	0.000	0.000
0.177	0.212	−0.000	0.711	−0.220	0.249	0.322	0.523	0.000
−0.000	0.509	0.508	0.389	−0.136	0.000	1.360	0.652	−0.261
0.177	0.114	0.000	−0.125	−0.000	0.000	−0.810	−0.442	−0.000
−0.000	0.136	0.217	−0.000	−0.000	0.859	0.497	0.350	−0.154
0.702	0.782	0.939	−0.385	−0.593	0.983	0.637	1.013	−0.781
0.789	0.866	1.019	−0.652	−0.000	1.571	0.497	0.957	−0.649
0.000	0.949	−0.000	−0.391	−0.000	2.394	3.004	2.922	−1.879
0.472	−0.000	−0.241	−0.282	−0.000	−0.890	0.000	0.000	0.378
−0.163	0.000	0.165	0.000	0.669	0.000	−0.000	−0.000	0.168
0.720	0.999	0.894	−0.603	−0.000	2.637	1.319	1.055	−0.752
−1.686	1.752	−1.195	0.000	0.000	0.965	0.797	0.480	−0.293
0.816	−0.711	0.242	−0.000	0.204	0.536	0.873	0.509	−0.287
−0.000	0.000	−0.000	−0.119	−0.266	0.490	0.146	−0.000	−0.000
−0.484	1.155	−0.907	0.000	0.000	0.847	0.288	−0.000	−0.294
−0.000	0.361	−0.530	−0.195	0.700	0.000	0.284	−0.000	0.000
0.530	−0.000	0.000	0.014	−0.060	0.753	0.686	0.406	0.501
−0.810	0.862	0.000	0.407	−0.512	−0.335	0.990	0.855	0.897
1.520	−0.828	0.485	−0.000	−0.392	0.682	−0.157	−0.117	0.374
0.153	0.250	0.185	−0.333	−0.440	0.584	0.831	0.304	0.732
0.750	0.132	0.521	−0.649	0.467	−0.697	0.653	0.509	0.765
0.578	0.159	0.470	0.589	1.294	−0.903	0.704	−0.000	0.444
0.000	−0.000	0.000	−0.256	−0.585	1.020	0.685	1.115	1.057
−0.000	0.707	0.732	0.293	0.330	0.000	0.583	−0.000	0.444
1.217	1.430	1.968	0.588	−0.560	0.896	1.242	−0.696	0.858
0.329	0.152	0.000	−0.774	0.634	−1.007	0.773	−0.682	0.800
0.393	0.000	0.348	0.411	−0.211	−0.000	0.688	−0.799	0.718
−0.681	−0.673	0.740	−0.431	0.706	−0.628	0.693	0.611	0.588
0.868	0.971	0.849	−0.314	−0.000	0.000	1.220	2.100	−1.887
0.000	0.534	0.716	−0.941	−0.761	1.008			
1.349	1.472	0.000	−1.027	−0.684	0.586			

Table A.3 Output matrix in the disturbance point $B_{2d}^{\mathrm{T}} \in \mathbb{R}^{1\times 148}$

Columns from 1 to 50	Columns from 51 to 100	Columns from 101 to 148
0.129	1.067	−1.276
0.212	0.939	−0.000
−0.424	0.989	0.862
−0.190	−0.000	0.633
0.416	0.924	1.132
0.162	0.000	1.455
0.836	0.320	0.733
−0.544	1.298	0.984
−1.085	0.578	0.000
0.175	−0.277	−0.359
0.392	−0.813	−0.000
0.333	−0.270	1.318
0.183	0.000	3.446
0.000	−0.833	−0.742
−0.000	1.322	0.000
0.000	−0.000	0.961
0.333	0.277	−0.704
−0.176	0.727	−0.331
0.577	0.804	0.444
0.762	0.874	0.206
0.330	−0.000	−0.480
−0.000	0.000	−0.974
0.270	0.603	1.135
−0.000	0.290	−0.288
0.417	0.673	−0.234
0.717	−0.350	0.000
−0.697	−1.182	−2.586
0.214	1.107	0.000
−0.000	−0.000	0.000
1.094	−2.032	−0.817
1.286	0.000	−0.569
−0.243	−0.361	−0.844
0.384	−0.137	0.540
1.195	−0.127	0.590
0.957	0.919	0.000
−0.000	−0.000	0.888
1.365	0.895	1.278
−0.000	−0.605	−0.190
0.368	0.527	0.921
0.288	0.810	1.144
0.237	−0.575	0.523
0.306	−0.090	1.896
0.000	0.922	0.185
1.488	−0.632	1.014
0.748	1.058	0.886
0.324	0.479	0.927
0.441	−0.321	1.169
−0.792	0.414	−1.336
0.382	1.335	
0.000	2.288	

Table A.4 Technical characteristics of the piezo actuators

Specifications	
Manufacturer	ACX
Model	QP20n QuickPack
Application type	strain actuator or bimorph actuator
Device size [mm]	$50.8 \times 25.4 \times 0.8$
Device weight [g]	4.82
Active elements	1 stack of 2 piezos
Piezo wafer size [mm]	$45.9 \times 20.6 \times 0.3$
Device capacitance [μF]	0.12
Full-scale voltage range [V]	±200

Table A.5 Technical characteristics of the power amplifiers

Specifications	
Manufacturer	ACX
Model	1224/5 QuickPack
Voltage gain [V/V]	−1 to −20
Maximum output power [WRMS]	20 (±200 V @ 200 mA peak)
Output impedance [Ω]	1 max.
Input impedance	4.4 to 104 kΩ (Voltage gain = 1 to 20)//20 pF
Input range [V]	±200 to ±10 (Voltage gain = 1 to 20)
Input fuse [mA]	5
Bandwidth [Hz]	5000
Nonlinearity	0.01% max.
Noise [VRMS]	0.001 typical
Load Impedance	5 μF max. @ 200 mA, 1 μF @ 50 mA
Power consumption [W]	100 max.
Weight [kg]	4.54

Table A.6 Technical characteristics of the accelerometers

Specifications	
Manufacturer	PCB Piezotronics
Model	M355B04
Sensing element	Ceramic
Sensitivity [mV/m/s^2]	100
Frequency range ±10% [Hz]	[0.6, 12.000]
Resonant frequency [kHz]	≥ 30
Amplitude range [m/s^2]	±50 peak
Resolution [m/s^2]	0.001
Mechanical shock limit [m/s^2]	50.000
Temperature operativity range [°C]	−53 to +93
Amplitude linearity	±1%
Transverse sensitivity	$\leq 3\%$
Excitation voltage [V]	18 to 30
Constant current excitation [mA]	2 to 8
Output impedance [Ω]	< 1000
Output bias voltage [V]	8 to 12
Discharge time constant [s]	0.5 to 2
Device weight [g]	11.2

Index

Actuator, 48, 55, 80, 82, 123, 133, 136, 143, 144, 155, 176
  conventional, 123, 124, 135
  electrodynamic, 124
  hydraulic, 124
  magnetostrictive, 127, 129, 130, 134, 139
  piezoelectric, 68, 79, 128, 144, 146, 153, 155, 163, 168, 187, 188, 192
  pneumatic, 124
  proof mass, 124
  selection, 123
  smart, 123, 129, 134, 180
Admissible functions, 19
Assumed modes, 18, 20

Bandpass controller, 84
Black-box approach, 30, 161
Boundary conditions
  geometric, 17
  natural, 17
Boundary value problem, 8–10, 12, 13, 15, 17, 19–21, 183
Bump noise, 3, 134

Cantilever beam, 15
Colocated measurements, 43, 108
Comparison functions, 19
Complete set, 19
Control, 183
  $\mathscr{H}_2$, 81, 91, 93, 94, 153, 166–168, 184
  $\mathscr{H}_\infty$, 80, 81, 83, 86, 87, 90, 93, 153, 166–168, 176, 179, 184
  active, 80, 129, 163
  actuator, 141
  adaptive strategy, 63, 65
  central $\mathscr{H}_\infty$, 90
  colocated, 169, 171
  energy, 170, 173, 175
  high gain, 61
  internal model, 58, 61
  LQG, 142, 143, 178
  modal, 58
  model-following, 135, 141, 176
  non-colocated, 169
  optimal, 58, 80, 165
  performance, 165, 169
  procedure, 153
  rectangular, 169, 176
  self-sensing, 144
  strongly stabilising, 80, 87
  symmetric output feedback, 91
Controllability Gramian, 96, 99, 113, 115
Controllability matrices, 33

Damping coefficient, 21, 25, 31, 40, 46, 52, 82, 156, 187
Damping matrix, 21
Differential eigenvalue problem, 8, 10–13, 16, 17, 21
Disturbance
  band-limited, 61
  broadband, 57, 67, 79, 153
  narrow-band, 72
  rejection, 61
  spectrum, 56, 171
  tonal, 57, 67

Eigenfunctions, 10, 19
Eigenvalue problem, 24
Eigenvalues, 10, 40
Eigenvectors, 22
Energy reduction index, 164
Everett integrals, 131, 132
Expansion theorem, 11

Finite element model, 18, 20
Force distribution, 25
Forgetting factors, 66

Generalised coordinates, 20
Generalised force distribution, 9
Generalised forces, 20, 22, 25
Gramian equation, 96
Gramians, 113, 169
Grey-box
  approach, 30, 31, 38, 41, 162
  identification, 22

Hankel matrix, 33
Hysteresis, 126, 127, 129
  compensation, 130, 132–134
  compensator, 132, 133
  model, 130, 132
  operator, 130, 133

Identification
  data, 49, 50
  dynamic matrix, 41
  exprimental, 179
  frequency domain, 29, 30, 34, 35, 141, 143
  problem, 31, 35, 38, 47, 133, 158
  procedure, 8, 20, 29–31, 34, 39, 40, 45–47, 52, 55, 131, 133, 141, 143, 153, 155, 179, 183
  results, 159
  subspace method, 30, 31, 33, 34, 38, 159
  time domain, 29, 30, 32, 48
Internal model principle, 62, 64, 65, 91

Least-squares technique, 37, 43
LMI, 81, 106, 154, 165, 168
Lueg patent, 3, 55

Magnetostrictive material, 127–129, 135, 176
Markov parameters, 33, 36
Mass coefficients, 20
Mass density, 21
Mass distribution, 9, 11
Mass matrix, 20, 22, 24, 47
Modal analysis, 8, 12, 31, 38, 39, 45–48, 51, 157, 162
Modal coordinates, 25, 38, 68, 82, 187
Modal costants, 47
Modal density, 56, 57, 75, 79, 80, 112
Modal equations, 12
Modal forces, 12, 82
Modal performance weights, 169, 171, 172, 174
Mode shapes, 1, 22, 23, 46, 52

Model
  actuator, 137
  continuous-time, 30, 41
  discrete-time, 30
  dynamic, 187
  finite-dimensional, 17, 19, 21, 22, 25, 58
  grey-box, 38
  order reduction, 40, 43, 47
Modes selection, 104

Natural frequency, 1, 8, 12, 14, 17, 23, 31, 39, 40, 44, 46, 52, 67, 68, 82, 154, 156, 160, 163, 172, 177, 187
Natural modes, 8, 12, 81, 84, 154, 166
Non-colocated measurements, 45, 108
Normal coordinates, 12
Normal mode shapes, 25
Normal modes, 12, 15

Observability Gramian, 113, 115
Observability matrices, 32
Operator
  adjoint, 8
  damping, 21
  linear, 9
  positive semi-definite, 9
  self-adjoint, 9, 20, 21
  stiffness, 9, 21

Piezoelectric material, 127
Preisach model, 130

Rapid control prototyping, 146, 148
Reciprocity property, 27

Sampled-data system, 39, 41
Scaling matrix, 98, 101
Sensor, 48, 55, 69, 80, 82, 128, 144, 153, 155
  accelerometer, 79, 155, 163
  FBG, 135, 139
  velocity, 67, 74
Separation principle, 60
Sliding manifold, 69
Smart auxiliary mass damper, 134–136, 155, 176
Smart material, 126, 127, 129, 176
Smart structure, 126
Spillover, 58–60, 75, 80, 176
  control, 59, 60
  observation, 59, 60
Standing wave, 22
State controller, 59
State feedback, 74
State observer, 59

Stiffness coefficients, 20
Stiffness matrix, 20, 22, 24

Test article
  "H"-shaped structure, 51
  fuselage stiffened skin panel, 153
Toeplitz matrix, 36
Travelling wave, 22

Uniform beam, 67, 72
Uniform string, 13

Youla–Kucera parametrisation, 63, 64

Other titles published in this series (continued):

Soft Sensors for Monitoring and Control of Industrial Processes
Luigi Fortuna, Salvatore Graziani, Alessandro Rizzo and Maria G. Xibilia

Adaptive Voltage Control in Power Systems
Giuseppe Fusco and Mario Russo

Advanced Control of Industrial Processes
Piotr Tatjewski

Process Control Performance Assessment
Andrzej W. Ordys, Damien Uduehi and Michael A. Johnson (Eds.)

Modelling and Analysis of Hybrid Supervisory Systems
Emilia Villani, Paulo E. Miyagi and Robert Valette

Process Control
Jie Bao and Peter L. Lee

Distributed Embedded Control Systems
Matjaž Colnarič, Domen Verber and Wolfgang A. Halang

Precision Motion Control (2nd Ed.)
Tan Kok Kiong, Lee Tong Heng and Huang Sunan

Optimal Control of Wind Energy Systems
Iulian Munteanu, Antoneta Iuliana Bratcu, Nicolaos-Antonio Cutululis and Emil Ceangă

Identification of Continuous-time Models from Sampled Data
Hugues Garnier and Liuping Wang (Eds.)

Model-based Process Supervision
Arun K. Samantaray and Belkacem Bouamama

Diagnosis of Process Nonlinearities and Valve Stiction
M.A.A. Shoukat Choudhury, Sirish L. Shah, and Nina F. Thornhill

Magnetic Control of Tokamak Plasmas
Marco Ariola and Alfredo Pironti

Real-time Iterative Learning Control
Jian-Xin Xu, Sanjib K. Panda and Tong H. Lee

Deadlock Resolution in Automated Manufacturing Systems
ZhiWu Li and MengChu Zhou

Model Predictive Control Design and Implementation Using MATLAB®
Liuping Wang

Fault-tolerant Flight Control and Guidance Systems
Guillaume Ducard

Predictive Functional Control
Jacques Richalet and Donal O'Donovan

Fault-tolerant Control Systems
Hassan Noura, Didier Theilliol, Jean-Christophe Ponsart and Abbas Chamseddine

Control of Ships and Underwater Vehicles
Khac Duc Do and Jie Pan

Detection and Diagnosis of Stiction in Control Loops
Mohieddine Jelali and Biao Huang

Stochastic Distribution Control System Design
Lei Guo and Hong Wang

Dry Clutch Control for Automotive Applications
Pietro J. Dolcini, Carlos Canudas-de-Wit, and Hubert Béchart

Nonlinear and Adaptive Control Design for Induction Motors
Riccardo Marino, Patrizio Tomei and Cristiano M. Verrelli
Publication due August 2010

Active Braking Control Design for Road Vehicles
Sergio M. Savaresi and Mara Tanelli
Publication due September 2010

MIX
Papier aus verantwortungsvollen Quellen
Paper from responsible sources
FSC® C105338
FSC
www.fsc.org

If you have any concerns about our products,
you can contact us on
ProductSafety@springernature.com

In case Publisher is established outside the EU,
the EU authorized representative is:
Springer Nature Customer Service Center GmbH
Europaplatz 3, 69115 Heidelberg, Germany

Printed by Libri Plureos GmbH
in Hamburg, Germany